# BIDDING AND TENDERING
# WHAT IS THE LAW?

### Third Edition

**Paul Sandori**

**William M. Pigott**

Bidding and Tendering: What Is The Law?, Third Edition
© LexisNexis Canada Inc. 2004
May 2004

All rights reserved. No part of this publication may be reproduced, stored in any material form (including photocopying or storing it in any medium by electronic means and whether or not transiently or incidentally to some other use of this publication) without the written permission of the copyright holder except in accordance with the provisions of the Copyright Act. Applications for the copyright holder's written permission to reproduce any part of this publication should be addressed to the publisher.

Warning: The doing of an unauthorized act in relation to a copyrighted work may result in both a civil claim for damages and criminal prosecution.

## Members of the LexisNexis Group worldwide

| | |
|---|---|
| Canada | LexisNexis Canada Inc, 75 Clegg Road, MARKHAM, Ontario |
| Argentina | Abeledo Perrot, Jurisprudencia Argentina and Depalma, BUENOS AIRES |
| Australia | Butterworths, a Division of Reed International Books Australia Pty Ltd, CHATSWOOD, New South Wales |
| Austria | ARD Betriebsdienst and Verlag Orac, VIENNA |
| Chile | Publitecsa and Conosur Ltda, SANTIAGO DE CHILE |
| Czech Republic | Orac sro, PRAGUE |
| France | Éditions du Juris-Classeur SA, PARIS |
| Hong Kong | Butterworths Asia (Hong Kong), HONG KONG |
| Hungary | Hvg Orac, BUDAPEST |
| India | Butterworths India, NEW DELHI |
| Ireland | Butterworths (Ireland) Ltd, DUBLIN |
| Italy | Giuffré, MILAN |
| Malaysia | Malayan Law Journal Sdn Bhd, KUALA LUMPUR |
| New Zealand | Butterworths of New Zealand, WELLINGTON |
| Poland | Wydawnictwa Prawnicze PWN, WARSAW |
| Singapore | Butterworths Asia, SINGAPORE |
| South Africa | Butterworth Publishers (Pty) Ltd, DURBAN |
| Switzerland | Stämpfli Verlag AG, BERNE |
| United Kingdom | Butterworths Tolley, a Division of Reed Elsevier (UK), LONDON, WC2A |
| USA | LexisNexis, DAYTON, Ohio |

**National Library of Canada Cataloguing in Publication**

Sandori, Paul, 1937-
   Bidding and tendering: what is the law? / Paul Sandori, William M. Pigott. — 3rd ed.

ISBN 0-433-43943-2

     1. Construction contracts—Canada. 2. Construction contracts—Canada—Cases. 3. Letting of contracts—Canada. 4. Letting of contracts—Canada—Cases. I. Pigott, William M. (William McCrea), 1942- . II. Title.

KE933.S25 2004                  343.71'078624                C2004-901439-0
KF902.S25 2004

# Dedication

This book is dedicated to The Right Honourable Madam Justice Beverley McLachlin, Chief Justice of Canada, a long-standing and much-appreciated member of the editorial board of the Construction Law Letter, where many of the comments and case summaries in this book first appeared.

This book is also dedicated to the late Right Honourable Mister Justice Willard Estey who wrote the judgment of the Supreme Court of Canada in *Ron Engineering* and revolutionized the law of bidding in Canada. Without his judgment, this book would not have seen the light of day.

# About the Authors

Paul Sandori is Senior Consultant with Revay and Associates Limited, Construction Consultants, and Editor of the *Construction Law Letter*. A member of the Ontario Association of Architects and Fellow of the Royal Architectural Institute of Canada, he has practised architecture in Europe and Canada for more than 25 years. Since joining Revay in 1992, he has acted as expert witness, mediator or arbitrator in numerous construction disputes, including bidding and tendering. He is the author of a book on structural mechanics, as well as many articles and papers on technical issues in construction.

Bill Pigott is with the Toronto office of Miller Thomson. His construction law practice begins at the "front end" of a project and covers a broad range of construction issues including provision for early claim resolution through negotiation/mediation. With respect to project execution plans, Bill advises on available options as to how the design and construction services will be procured, including expressions of interest, binding procurement, requests for proposal and conventional bids. Once the execution plan has been established, he prepares and negotiates contracts using amended versions of standard industry contracts where appropriate. Bill's experience in project modelling runs from general contracts based on competitively bid drawings and specifications to more demanding forms of project execution including construction management, design build and public/private partnerships. His experience covers most areas of the construction industry including power generation and health care. While much of Bill's work is for owners and their consultants, he understands the contractor's side of the equation and provides advice in connection with contract negotiation, bid issues and claims resolution. Bill is a "contractor's kid".

# Preface to the Third Edition

The second edition of this book was published in April 2000. Since then, the Supreme Court of Canada has twice addressed the core findings in the decision that revolutionized the law of bidding — *Ron Engineering* — and affirmed them. The Court also expanded the reach of *Ron Engineering* by addressing fairness and by explaining how Contract A works at the contractor/subcontractor level. Lower courts have been busy, too.

As authors, we have two goals for this third edition.

First, we intend to update the book for our readers. The need is obvious — there are important new court decisions. In the process of updating, we have tried to refine our thinking in the second edition.

Second, we hope to encourage the industry to change its mindset on bidding, and to reflect that changed thinking in the language of the bid process. *Ron Engineering* created a new paradigm yet much of the industry still operates on the old model — language and all. Switching to the new model should result in a less adversarial bidding process, and less litigation.

In keeping with the example set in the second edition, we must counsel our readers that the third edition is still no substitute for your lawyer. We do not presume to give you legal advice.

Paul Sandori<br>
William M. Pigott<br>
April 2004

# Preface to the Second Edition

Our common law is made by judges. In this publication, which has now been through five revisions, we bring you the key decisions which represent the law of competitive bidding in Canada.

Almost 20 years ago, the Supreme Court of Canada started a revolution in the law as it affects bidding: it created the Contract A/Contract B approach which has deeply affected the rights and duties of owners, contractors and subcontractors. In this publication, we track the development of the law and explain these rights and duties.

The law is alive; it is constantly changing. Some decisions conflict. Some areas of the law are not settled. But you cannot allow this seeming instability to put you off; you must keep up to date. Some of the cases that follow are a good illustration of what happens if you don't.

In one important respect, this text is different from all other books on the law of bidding and tendering: we do not give you just the legal principles but also a detailed summary of each case. Thus, you find out not only what the outcome of a bidding dispute in the courts was, but you also find out what caused the dispute or problem in the first place, how it developed and what the participants did, or did not do, to end up in court.

You learn from the mistakes of others so that if confronted by a similar situation, you can choose a different path, one that leads to a settlement rather than to the courtroom.

Of course, there is no guarantee that you can predict the outcome of a dispute. Litigation is like baseball. In baseball, "the game ain't over till the fat lady (or gentleman) sings". In litigation, you cannot know the outcome before the judge reads it. The tune in the end may come as a total surprise, and sometimes experienced lawyers are just as amazed as people in the construction industry.

Finally, this would not be a publication touching on legal issues if we did not include a disclaimer: the cases printed here are *no substitute for your lawyer* — we would not presume to give you legal advice.

Neither are we trying to turn you into a lawyer. The most important objective of this book is to keep you out of the legal morass in which so many other construction industry participants seem to get lost.

Paul Sandori<br>
William M. Pigott<br>
Toronto, Ontario<br>
April 2000

# ACKNOWLEDGMENTS

Book writing is a team sport. The authors want to recognize our team members whose names would otherwise not find their way into this book.

Miller Thomson LLP and Revay and Associates Limited provided the infrastructure and the help of several members of its staff. The authors thank in particular, Brenda Shepperdson for her patience and hard work. The authors also thank the members of the word processing department of Miller Thomson LLP who typed and retyped the manuscript – ad nauseam.

The authors are indebted to the research department of Miller Thomson LLP and, in particular its leader, Innes Freeman and her colleague Dian-Marie Galita whose research skills and insights helped track down case law and other sources of support for our work.

Finally, we each thank our families who have supported the writing of this book and who endured us with grace in the process.

# TABLE OF CONTENTS

*Preface to the Third Edition* ..................................................................... vii
*Preface to the Second Edition* ..................................................................... ix
*Acknowledgments* .......................................................................................... xi
*Table of Cases* ............................................................................................... xix

**Chapter 1: Traditional Law of Bidding** ............................................................. 1
    Dickinson v. Dodds ............................................................................................. 1
    McMaster University v. Wilchar Construction Ltd. ........................................ 3
    Belle River Community Arena Inc. v. W.J.C. Kaufmann Co. Ltd ................ 4
    Summary ............................................................................................................. 6

**Chapter 2: Revolution in the Law of Bidding** .................................................. 9
    R. v. Ron Engineering & Construction (Eastern) Ltd. ................................. 10
    Calgary (City) v. Northern Construction Company Ltd. ............................. 14
    Summary ........................................................................................................... 17

**Chapter 3: Duties and Rights of Owners** ....................................................... 19
    Best Cleaners and Contractors Ltd. v. R. ...................................................... 19
    Elgin Construction Ltd. v. Russell Township ................................................ 21
    Chinook Aggregates Ltd. v. Abbotsford (Municipal District) ..................... 23
    Acme Building and Construction Limited v. Newcastle (Town) ................. 25
    Megatech Contracting Limited v. Ottawa-Carleton (Regional
        Municipality) ............................................................................................. 26
    Kencor Holdings Ltd. v. Saskatchewan ......................................................... 28
    Power Agencies Co. Ltd. v. Newfoundland Hospital and Nursing
        Home Association ..................................................................................... 30
    Martselos Services Limited v. Arctic College ............................................... 32
    Colautti Brothers Marble Tile and Carpet (1985) Inc. v. Windsor
        (City) ........................................................................................................... 35
    Thompson Bros. (Const.) Ltd. v. Wetaskiwin (City) ..................................... 39
    M.J.B. Enterprises Ltd. v. Defence Construction (1951) Ltd. ..................... 41
    Tarmac Canada Inc. v. Hamilton-Wentworth (Regional
        Municipality) ............................................................................................. 46
    Sound Contracting Ltd. v. City of Nanaimo ................................................. 49
    Midwest Management (1987) Ltd. v. B.C. Gas Utility Ltd. ......................... 52
    Martel Building Ltd. v. Canada ..................................................................... 57
    Cable Assembly Systems Ltd. v. Dufferin-Peel Roman Catholic
        School Board ............................................................................................. 62
    J. Oviatt Contracting Ltd. v. Kitimat General Hospital Society ................. 66
    Wind Power Inc. v. Saskatchewan Power Corp. ........................................... 68

Mellco Developments Ltd. v. Portage la Prairie (City) ............................ 72
Kinetic Construction Ltd. v. Comox-Strathcona (Regional District) ........... 76
Summary ........................................................................... 79

**Chapter 4: Subcontractor Bids** ........................................................ 81
Peddlesden Ltd. v. Liddell Construction Ltd. ........................................ 81
Westgate Mechanical Contractors Ltd. v. PCL Construction Ltd. ................. 84
Ron Brown Ltd. v. Johanson ........................................................ 86
Bate Equipment Ltd. v. Ellis-Don Limited .......................................... 88
Scott Steel (Ottawa) Ltd. v. R.J. Nicol Construction (1975) Ltd. ................... 91
Vipond Automatic Sprinkler Co. Ltd. v. E.S. Fox Ltd. .............................. 95
Stuart Olson Constructors v. NAP Building Products ............................. 98
Naylor Group Inc. v. Ellis-Don Construction Ltd. ................................. 100
Summary .......................................................................... 106

**Chapter 5: Supplier Bids** ............................................................. 107
Arrow Construction Products Ltd. v. Nova Scotia
   (Attorney General) ............................................................. 107
Western Plumbing and Heating Ltd. v. Industrial Boiler-Tech Inc. ............. 111
Summary .......................................................................... 115

**Chapter 6: Mistake in Bid** ........................................................... 117
Gloge Heating & Plumbing Ltd. v. Northern Const. Company Ltd. .............. 117
Town of Vaughan v. Alta Surety Company ........................................ 119
Forest Contract Management v. C&M Elevator Ltd. .............................. 121
Vachon Construction Ltd. v. Cariboo (Regional district) ........................ 123
Vic Van Isle Construction Ltd. v. School District No. 23 ......................... 127
North Vancouver (District) v. Progressive Contracting
   (Langley) Ltd. .................................................................. 131
City of Ottawa Non-Profit Housing Corp. v. Canvar Construction
   (1991) Inc. ..................................................................... 134
Derby Holdings Ltd. v. Wright Construction Western Inc. ...................... 138
Graham Industrial Services Ltd. v. Greater Vancouver Water
   District ........................................................................ 142
Toronto Transit Commission v. Gottardo Construction Limited ............... 146
Summary .......................................................................... 150

**Chapter 7: Negligence and Misrepresentation** ...................................... 151
Hedley Byrne & Co. Ltd. v. Heller & Partners Ltd. ................................ 151
Carman Construction Ltd. v. Canadian Pacific Railway Co. ..................... 154
Mawson Gage Associates Ltd. v. R. ............................................... 157
St. Lawrence Cement Inc. v. Farry Grading & Excavating Limited ............ 159
Opron Construction Co. Ltd. v. R. ................................................ 161

BG Checo International Limited v. B.C. Hydro and Power
    Authority ................................................................................ 169
Twin City Mechanical v. Bradsil (1967) Limited ...................... 173
Ken Toby Ltd. v. B.C. Buildings Corporation ............................ 175
Summary ..................................................................................... 179

**Chapter 8: Duties of Consultants** ................................................ 181
Cardinal Construction Ltd. v. Brockville (City) ......................... 181
Brown & Huston Limited v. York (Borough) ............................ 184
Edgeworth Construction Ltd. v. N.D. Lea & Associates Ltd. .... 186
Auto Concrete Curb Ltd. v. South Nation River Conservation
    Authority ................................................................................ 189
ACL Holdings Ltd. v. St. Joseph's Hospital of Estevan ............. 191
Wigmar Construction (B.C.) Ltd. v. Defence Construction
    (1951) Limited ....................................................................... 193
J. P. Metal Masters Inc. v. David Mitchell Co. ........................... 196
WIB Co. Construction Ltd. v. School District No. 23
    (Central Okanagan) ................................................................ 198
JJM Construction Ltd. v. Sandspit Harbour Society .................. 202
Summary ..................................................................................... 207

**Chapter 9: Special Remedies** ....................................................... 209
Thomas C. Assaly Corporation Ltd. v. R. .................................. 209
Peter Kiewit Sons Co. v. Richmond (City) ................................ 212
Cegeco Construction Ltée v. Ouimet ......................................... 215
Ellis-Don Construction Ltd. v. Canada (Minister of Public Works) .......... 217
Donohue Recycling Inc. (c.o.b. Abitibi Consolidated Recycling Division)
    v. Toronto (City) .................................................................... 219
Summary ..................................................................................... 222

**Chapter 10: Time of Bid Submittal** ............................................. 223
Smith Bros. & Wilson (B.C.) Ltd. v. B.C. Hydro ....................... 223
Bradscot (MCL) Ltd. v. Hamilton-Wentworth Catholic School
    Board ...................................................................................... 225
Summary ..................................................................................... 228

**Chapter 11: Bid Compliance and Bid Evaluation** ..................... 229
Introductory Note ....................................................................... 229
Compliance ................................................................................. 230
    Owner Obligations/Owner Control ....................................... 231
    When Does Contract A Not Arise? ........................................ 232
    Contractors/Subcontractors ................................................... 234
Evaluation ................................................................................... 236
    Privilege Clause ..................................................................... 236

Fairness in Bidding .................................................................................... 240
Privilege Clause/Discretion Clause ........................................................ 244

**Chapter 12: Troublesome Bid Issues** ........................................................ 247
   Compliance Issues ................................................................................... 247
      On-Time Bid Submission ................................................................. 247
      Is This A Valid Offer? ....................................................................... 248
      Plans and Specs? ............................................................................... 248
      Bid Security/Agreement to Bond ..................................................... 249
      Failure to Nominate Subcontractors ............................................... 250
      Bidder Error in Bid: Latent Mistake Revisited ............................... 250
      Substantial Compliance/Irregular Bids Revisited .......................... 251
      All Bids Are Non-Compliant ............................................................ 253
         (a) What Does The Owner Have? ................................................ 254
         (b) Evaluation: What Is An Owner To Do? ................................ 254
         (c) Reject First and Then Negotiate? ........................................... 255
         (d) First Principles ......................................................................... 255
   Evaluation Issues ..................................................................................... 256
      Pre-Qualification ............................................................................... 257
      Nuanced View of Cost ...................................................................... 258
      Unrealistically Low Bids .................................................................. 259
      Unbalanced Bids ................................................................................ 260
      After-Bid Amendments By Contractors ......................................... 261
      After-Bid Amendments By Owners ................................................ 261
      Unsolicited Alternatives ................................................................... 263
      Owner Errors/Omissions in Bid Documents ................................. 264
      Subcontractor Error in Bid ............................................................... 266
      Tied Bids ............................................................................................ 268
      Electronic Bidding ............................................................................. 268
   Where To From Here? ............................................................................. 269

**Chapter 13: Damages and Other Relief** ................................................... 271
   General Principles ................................................................................... 271
   Proof of Damages .................................................................................... 273
   Breach of Contract A ............................................................................... 275
      Awards In Favour of Owners Against Contractors ....................... 275
      Awards In Favour of Contractors Against Owners ...................... 276
      Awards In Favour of Subcontractors Against Contractors .......... 280
      Awards In Favour of Contractors Against Subcontractors .......... 282
   Misrepresentation .................................................................................... 283
      Awards In Favour of Subcontractors Against Owners ................. 283
      Contractors Against Owners ............................................................ 284
   Other Relief .............................................................................................. 285
      Unjust Enrichment ............................................................................ 285

Prerogative Remedies/Injunction..........................................................286
Afterthought ............................................................................................288

**Chapter 14: Contract A: Treat It Like A Contract**....................289
   Contract A Is A Real Contract.................................................................289
   Treat Contract A Like a Contract............................................................290
      New Law/New Practice .....................................................................290
      Identifying and Allocating Risks .......................................................291
      Keep It Simple — Keep Your Options Open ...................................292
      Dispute Resolution............................................................................294
   So, Where Are We?.................................................................................295

**Chapter 15: Conclusion** ................................................................297

# TABLE OF CASES

## A

ACL Holdings Ltd. v. St. Joseph's Hospital of Estevan, [1996] 6 W.W.R. 207, 29 C.L.R. (2d) 88 (Sask. Q.B.) .................................................................. 191, 193
Acme Building and Construction Limited v. Newcastle (Town) (1990), 38 C.L.R. 56 (Ont. Dist. Ct.), affd (1992), 2 C.L.R. (2d) 308 (Ont. C.A.), leave to appeal refd (1993), 151 N.R. 394n (S.C.C.)
.................................................................. 25, 31, 37, 40, 47, 49, 258, 261, 276
Arkwright v. Newbold (1881), 17 Ch.D. 301 (Eng. C.A.) ............................................ 162
Arrow Construction Products Ltd. v. Nova Scotia (Attorney General) (1995), 145 N.S.R. (2d) 336, 418 A.P.R. 336 (S.C.), revd (1996), 27 C.L.R. (2d) 1, 150 N.S.R. (2d) 241, 436 A.P.R. 241 (C.A.), leave to appeal refd (1997), 461 A.P.R. 320n, 208 N.R. 244n (S.C.C.) .................................................................. 107, 194
Auto Concrete Curb Ltd. v. South Nation River Conservation Authority (1988), 30 C.L.R. 245 (Ont. H.C.), affd (1992), 2 C.L.R. (2d) 262, 89 D.L.R. (4th) 393 (Ont. C.A.), additional reasons (1992), 89 D.L.R. (4th) 403, 2 C.L.R. (2d) 262 at 272 (Ont. C.A.), revd, [1993] 3 S.C.R. 201, 105 D.L.R. (4th) 382, 12 C.L.R. (2d) 190 .................................................................. 189, 264, 265

## B

BG Checo International Limited v. B.C. Hydro and Power Authority, [1990] 3 W.W.R. 690, 41 C.L.R. 1 (B.C.C.A.), affd, [1993] 1 S.C.R. 12, [1993] 2 W.W.R. 321, 5 C.L.R. (2d) 173, application for re-hearing refd (1993), 5 C.L.R. (2d) 173n (S.C.C.) .................................................................. 169, 265, 273, 274, 284
Banque de Montréal v. Hydro-Québec, [1990] R.R.A. 3 (C.A. Qué), revd in part, [1992] 2 S.C.R. 554, 3 C.L.R. (2d) 1 .................................................................. 163
Bate Equipment Ltd. v. Ellis-Don Limited (1992), 2 C.L.R. (2d) 157, 132 A.R. 161 (Q.B.), affd (1993), 157 A.R. 274, 77 W.A.C. 274 (C.A.), leave to appeal refd (1995), 190 N.R. 398n, 135 W.A.C. 78n (S.C.C.) .................................................................. 88, 93
Belle River Community Arena Inc. v. W.J.C. Kaufmann Co. Ltd. (1977), 15 O.R. (2d) 738, 76 D.L.R. (3d) 629 (H.C.), affd (1978), 20 O.R. (2d) 447, 87 D.L.R. (3d) 761 (C.A.) .................................................................. 4, 9, 112, 254
Best Cleaners and Contractors Ltd. v. R., [1985] 2 F.C. 293, 58 N.R. 295 (C.A.) .................................................................. 19, 24, 31, 34, 47, 125, 214, 224, 238, 241, 242
Bradscot (MCL) Ltd. v. Hamilton-Wentworth Catholic School Board (1999), 42 O.R. (3d) 723, 44 C.L.R. (2d) 1 (C.A.) .................................................................. 225, 247
British Columbia v. SCI Engineers & Constructors Inc., [1993] B.C.J. No. 248, 22 B.C.A.C. 89 (C.A.) .................................................................. 67, 78, 251
Brown & Huston Limited v. York (Borough) (1983), 5 C.L.R. 240 (Ont. H.C.), vard (1985), 17 C.L.R. 192 (Ont. C.A.) .................................................................. 184, 196, 284

## C

Cable Assembly Systems Ltd. v. Dufferin-Peel Roman Catholic School Board, [2002] O.J. No. 379, 155 O.A.C. 139, 13 C.L.R. (3d) 163 (C.A.) .................................................................. 62

Calgary (City) v. Northern Construction Company Ltd. (1982), 23 Alta. L.R. (2d) 338 (Q.B.), revd (1985), 42 Alta. L.R. (2d) 1, [1986] 2 W.W.R. 426, 19 C.L.R. 287 (C.A.), affd, [1987] 2 S.C.R. 757, 56 Alta. L. R. (2d) 193, [1988] 2 W.W.R. 193 .................................... 14,16, 121, 128, 135, 136, 258, 261, 276

CanAmerican Auto Lease and Rental Ltd. v. R., unreported, March 4, 1985, Doc. No. T-4780-76 (F.T.D.), affd, [1987] 3 F.C. 144 (C.A.) ........................................................................................................................ 279, 280

Cardinal Construction Ltd. v. Brockville (City) (1984), 25 M.P.L.R. 116, 4 C.L.R. 149 (Ont. H.C.) ..................................................... 162, 168, 181, 204

Carman Construction Ltd. v. Canadian Pacific Railway Co. (1982), 136 D.L..R. (3d) 193, 18 B.L.R. 65 (S.C.C.) ........................................................ 154

Catre Industries Ltd. v. Alberta (1987), 97 A.R. 1 (Q.B.), revd (1989), 63 D.L.R. (4th) 74, 36 C.L.R. 169 (Alta. C.A.), leave to appeal refd (1990), 108 N.R. 170n (S.C.C.) .......................................................................... 164

Cegeco Construction Ltée v. Ouimet (1991), 50 C.L.R. 171, 48 F.T.R. 143 (T.D.) ............................................................................................ 215, 244

Central Trust Company v. Rafuse (*sub nom.* Central & Eastern Trust Co. v. Rafuse), [1986] 2 S.C.R. 147, 31 D.L.R. (4th) 481, 75 N.S.R. (2d) 109, vard, [1988] 1 S.C.R. 1206 ........................................................................ 160

Chinook Aggregates Ltd. v. Abbotsford (Municipal District) (1987), 28 C.L.R. 290 (B.C. Co. Ct.), affd (1989), [1990] 1 W.W.R. 624, 35 C.L.R. 241 (B.C.C.A.), [1990] B.C.W.L.D. 56 (B.C.C.A.) .................... 23, 24, 26, 29, 31, 34, 40, 47, 52, 125, 126, 178, 214, 224, 237, 239, 241, 264

Colautti Brothers Marble Tile and Carpet (1985) Inc. v. Windsor (City) (1996), 36 M.P.L.R. (2d) 258 (Ont. Gen. Div.) ............................................. 35, 277

Cornwall Gravel Co. Ltd. v. Purolator Courier Ltd. (1978), 18 O.R. (2d) 551, 83 D.L.R. (3d) 267 (H.C.), affd (1979), 28 O.R. (2d) 704, 115 D.L.R. (3d) 511 (C.A.), affd, [1980] 2 S.C.R. 118, 120 D.L.R. (3d) 575 .................. 271, 272, 273

# D

Derby Holdings Ltd. v. Wright Construction Western Inc., [2002] S.J. No. 367 (Q.B.), affd [2003] S.J. No. 588 (C.A.) .......................... 137, 138, 141, 142

Derry v. Peek (1887), 37 Ch. D. 541 (Eng. C.A.), revd [1889] All E.R. Rep. 1, 14 A.C. 337 (U.K. H.L.) .......................................................................... 165

Dickinson v. Dodds (1876), 2 Ch. D. 463 (Eng. Ch. Div.) .................................. 1, 12, 289

Donohue Recycling Inc. (c.o.b. Abitibi Consolidated Recycling Division) v. Toronto (City), [2002] O.J. No. 2785 (S.C.J.) ........................ 219, 222, 287

# E

Edgeworth Construction Ltd. v. N.D. Lea & Associates Ltd. (1989), 37 C.L.R. 152 (B.C.S.C.), affd (1991), 44 C.L.R. 88, [1991] 4 W.W.R. 251 (B.C.C.A.), revd in part, [1993] 3 S.C.R. 206, 107 D.L.R. (4th) 169, 12 C.L.R. (2d) 161 ................................ 163, 186, 189, 195, 197, 204, 206, 207, 264

Elgin Construction Ltd. v. Russell Township (1987), 24 C.L.R. 253 (Ont. H.C.) ............................................. 21, 23, 24, 31, 34, 40, 237, 241, 258

Elite Bailiff Services Ltd. v. British Columbia, [2003] B.C.J. No. 376, 2003 BCCA 102, 223 D.L.R. (4th) 39, [2003] 4 W.W.R. 228, 10 B.C.L.R. (4th) 264 (C.A.) .............................................................................................. 78

Ellis-Don Construction Ltd. v. Canada (Minister of Public Works) (1992), 1 C.L.R. (2d) 193, 54 F.T.R. 42 (T.D.) ........................... 28, 217, 244, 287

## F

Forest Contract Management v. C&M Elevator Ltd. (1988), 33 C.L.R. 118, 93 A.R. 388 (Q.B.) .................................................................................................. 121
Foundation Building West Inc. v. Vancouver (City) (1995), 22 C.L.R. (2d) 94, [1995] B.C.J. No. 861 (S.C.) ............................................................................... 67

## G

George Wimpey Canada Ltd. v. Hamilton-Wentworth (Regional Municipality) *See* Tarmac Canada Inc. v. Hamilton-Wentworth (Regional Municipality)
Glenview Corporation v. R. (1990), 34 F.T.R. 292 (T.D.) ........................................ 218
Gloge Heating & Plumbing Ltd. v. Northern Const. Company Ltd. (1984), 5 C.L.R. 212, [1984] 3 W.W.R. 63, 6 D.L.R. (4th) 450 (Alta. Q.B.), affd, [1986] 2 W.W.R. 649, 27 D.L.R. (4th) 264, 19 C.L.R. 281 (Alta. C.A.)
.................................................................................................... 99, 112, 117, 282
Graham Industrial Services Ltd. v. Greater Vancouver Water District, [2003] B.C.J. No. 2614, 2003 BCSC 1735 (S.C.), affd [2004] B.C.J. No. 5, 2004 BCCA 5 (C.A.) ............................................................. 79, 142, 144, 145, 246, 263

## H

Hadley v. Baxendale (1854), 156 E.R. 145, [1843-60] All E.R. Rep. 461 (Eng. Ex. Div.) ............................................................................................................. 272
Haggerty v. Victoria (City) (1895), 4 B.C.R. 163 (S.C.) ................................... 212, 214
Harry v. Kreutziger (1977), 3 B.C.L.R. 348 (S.C.), revd (1978), 95 D.L.R. (3d) 231, 9 B.C.L.R. 166 (C.A.) .................................................................................. 135
Hedley Byrne & Co. Ltd. v. Heller & Partners Ltd., [1961] 3 All E.R. 891 (Eng. C.A.), affd (1963), [1964] A.C. 465, [1963] 2 All E.R. 575 (U.K. H.L.) ............................. 109, 110, 151, 156, 160, 184, 187, 194, 202, 285
Heilbut, Symons & Co. v. Buckleton (1912), [1913] A.C. 30, [1911-13] All E.R. Rep. 83 (U.K. H.L.) ..................................................................................... 155
Hollerbach v. United States (1914), 233 U.S. 165 ......................................................... 168

## I

Inglis v. Buttery (1878), 3 A.C. 552 (U.K. H.L.) ........................................................ 129
Ip v. Insurance Corp. of British Columbia (1994), 89 B.C.L.R. (2d) 251, 21 C.C.L.I. (2d) 26, 23 C.P.C. (3d) 345 (S.C.) ........................................................... 201

## J

J. Oviatt Contracting Ltd. v. Kitimat General Hospital Society, [2002] B.C.J. No. 1254, 2002 BCCA 323, 171 B.C.A.C. 190, 16 C.L.R. (3d) 111 (C.A.) ........ 66, 78
JJM Construction Ltd. v. Sandspit Harbour Society (1998), 38 C.L.R. (2d) 179 (B.C.S.C.), affd (1999), 49 C.L.R. (2d) 149 (B.C.C.A.), additional reasons, January 31, 2000, [2000] B.C.J. No. 192, and February 4, 2000, [2000] B.C.J. No. 234, and March 27, 2000 [2000] B.C.J. No. 605 (B.C.C.A.) ............................... 202
J. P. Metal Masters Inc. v. David Mitchell Co., [1998] B.C.J. No. 551, 105 B.C.A.C. 95, 49 B.C.L.R. (3d) 88, 37 C.L.R. (2d) 1 (C.A.) ............................ 196, 198

## K

K.R.M. Construction Ltd. v. B.C. Railway (1981), 18 C.L.R. 159 at 169 (B.C.S.C.), revd (1982), 40 B.C.L.R. 1, 18 C.L.R. 159 at 277 (C.A.) ...................... 162

Ken Toby Ltd. v. B.C. Buildings Corporation, [1997] 8 W.W.R. 721, 34 C.L.R. (2d) 81 (B.C.S.C.), revd (1999), 45 C.L.R. (2d) 141, 173 D.L.R. (4th) 169 (B.C.C.A.), leave to appeal to S.C.C. dismissed without reasons, [1999] S.C.C.A. No. 263 ............................................................... 175, 179, 275, 283, 284

Kencor Holdings Ltd. v. Saskatchewan, [1991] 6 W.W.R. 717, 96 Sask. R. 171 (Q.B.) ........................................................................................... 28, 47, 241, 276

Kinetic Construction Ltd. v. Comox-Strathcona (Regional District), [2003] B.C.J. No. 2533, 2003 BCSC 1673 (S.C.) ...................................... 76, 144, 262

## L

Lac Minerals Ltd. v. International Corona Resources Ltd. (1986), 53 O.R. (2d) 737, 25 D.L.R. (4th) 504 (H.C.), affd (1987), 62 O.R. (2d) 1, 44 D.L.R. (4th) 592 (C.A.), affd, [1989] 2 S.C.R. 574, 61 D.L.R. (4th) 14 ............... 165

London County Council v. Henry Boot and Sons Ltd., [1959] 3 All E.R. 636 (U.K. H.L.) ............................................................................................. 133

Lumley v. Gye (1853), [1843-60] All E.R. Rep. 208, 118 E.R. 749 (Eng. Q.B.) .......... 192

## M

M.J.B. Enterprises Ltd. v. Defence Construction (1951) Ltd. (1994), 164 A.R. 399, 18 C.L.R. (2d) 120 (Q.B.), affd (1997), 196 A.R. 124, 33 C.L.R. (2d) 1 (C.A.), revd, [1999] 1 S.C.R. 619, 170 D.L.R. (4th) 577, 44 C.L.R. (2d) 163 ................................................................................. 16, 41, 50, 51, 52, 53, 55, 63, 65, 67, 68, 70, 72, 74, 75, 76, 229, 232, 234, 236, 237, 245, 258, 274, 289, 290

Martel Building Ltd. v. Canada, 2000 SCC 60, [2000] 2 S.C.R. 860, [2000] S.C.J. No. 60 ......................................................... 17, 57, 61, 70, 72, 77, 79, 229, 234, 242, 243, 244, 246, 269, 274, 290

Martselos Services Limited v. Arctic College (1992), 5 B.L.R. (2d) 204 (N.W.T.S.C.), revd, [1994] 3 W.W.R. 73, 12 C.L.R. (2d) 208, 111 D.L.R. (4th) 65 (N.W.T.C.A.), leave to appeal refd (1994), 17 C.L.R. (2d) 59n (S.C.C.) ........................................................................................ 32, 34, 242

Mawson Gage Associates Ltd. v. R. (1987), 13 F.T.R. 188 (T.D.) ............................................................... 157, 265, 266, 283, 286

McMaster University v. Wilchar Construction Ltd., [1971] 3 O.R. 801, 22 D.L.R. (3d) 9 (H.C.), affd (1973), 69 D.L.R. (3d) 400n (Ont. C.A.) ................................................................................ 3, 11, 118, 134, 251

Megatech Contracting Limited v. Ottawa-Carleton (Regional Munici-pality) (1989), 34 C.L.R. 35, 68 O.R. (2d) 503 (H.C.) ............................................................................ 26, 31, 250, 258

Mellco Developments Ltd. v. Portage la Prairie (City), [2001] M.J. No. 388, 2001 MBQB 236, [2001] 11 W.W.R. 282, 167 Man. R. (2d) 161, 21 B.L.R. (3d) 25, 11 C.L.R. (3d) 227, 22 M.P.L.R. (3d) 227 (Q.B.), affd [2002] M.J. No. 381, 2002 MBCA 125, 222 D.L.R. (4th) 67 (C.A.), leave to appeal to S.C.C. dismissed without reasons [2002] S.C.C.A. No. 502 ........................... 72, 233

Midwest Management (1987) Ltd. v. B.C. Gas Utility Ltd. (1999), 47 C.L.R. (2d) 101 (B.C.S.C.), vard (2000), 5 C.L.R. (3d) 140 (B.C.C.A.) ............................................. 52, 61, 79, 233, 246, 248, 249, 262, 263, 274

TABLE OF CASES **xxiii**

Morrison-Knudsen v. B.C. Hydro and Power Authority (1976), 112 D.L.R.
(3d) 397 (B.C.C.A.) ........................................................................................... 167

## N

Naylor Group Inc. v. Ellis-Don Construction Ltd. (1996), 13 O.T.C. 141, 30
C.L.R. (2d) 195 (Gen. Div.), revd (1999), 171 D.L.R. (4th) 243
(Ont. C.A.), vard [2001] 2 S.C.R. 943, [2001] S.C.J. No. 56, 2001
SCC 58 ..................... 17, 100, 229, 234, 235, 236, 267, 275, 280, 281, 285, 286, 290
North Vancouver (District) v. Progressive Contracting (Langley) Ltd. (1992),
49 C.L.R. 298 (B.C.S.C.), affd (1993), 78 B.C.L.R. (2d) 356, 6 C.L.R. (2d)
136 (C.A.) ........................................................................................................ 131

## O

Opron Construction Co. Ltd. v. R. (1994), 14 C.L.R. (2d) 97, 151 A.R. 241
(Q.B.) ............................................................................................................... 161
Ottawa (City of) Non-Profit Housing Corp. v. Canvar Construction (1991)
Inc., [1999] O.J. No. 1972 (S.C.J.), revd (2000), 3 C.L.R. (3d) 55, [2000]
O.J. No. 1078 (C.A.) ........................................................... 134, 137, 258, 276

## P

Peddlesden Ltd. v. Liddell Construction Ltd. (1981), 32 B.C.L.R. 392, 128
D.L.R. (3d) 360 (S.C.) ........................................................... 81, 89, 93, 94
Peter Kiewit Sons Co. v. Richmond (City) (1992), 11 M.P.L.R. (2d) 110, 1
C.L.R. (2d) 5 (B.C.S.C.) ...................................................... 212, 244, 287
Powder Mountain Resorts Ltd. v. British Columbia, [1999] 11 W.W.R. 168
(B.C.S.C.), affd [2001] B.C.J. No. 2172, [2001] 11 W.W.R. 168, 2001
BCCA 619 ........................................................................................................ 73
Power Agencies Co. Ltd. v. Newfoundland Hospital and Nursing Home
Association (1991), 90 Nfld. & P.E.I.R. 64, 280 A.P.R. 64, 44 C.L.R. 255
(Nfld. T.D.) ...................................................................................................... 30

## Q

Queen v. Cognos (1987), 63 O.R. (2d) 389, 18 C.C.E.L. 146 (H.C.), revd
(1990), 69 D.L.R. (4th) 288, 38 O.A.C. 180 (C.A.), revd, [1993] 1 S.C.R.
87, 99 D.L.R. (4th) 626 ..................................................... 109, 153, 158, 194

## R

R. v. Ron Engineering & Construction (Eastern) Ltd. (1979), 24 O.R. (2d) 332, 98
D.L.R. (3d) 548 (C.A.), revd, [1981] 1 S.C.R. 111, 119 D.L.R. (3d) 267
................................ 9, 10, 13, 14, 16, 17, 19, 28, 29, 30, 33, 35, 40, 41, 42, 45, 48, 52,
54, 55, 63, 65, 67, 68, 74, 75, 78, 82, 83, 85, 89, 92, 93, 101, 112,
115, 117, 118, 122, 124, 130, 135, 136, 137, 141, 150, 151, 152, 159,
175, 176, 213, 222, 224, 226, 227, 229, 231, 232, 234, 236, 240, 246,
248, 250, 251, 252, 253, 254, 256, 258, 265, 269, 274, 275, 289, 290, 297
Redgrave v. Hurd (1881), 20 Ch. D. 1, [1881-5] All E.R. Rep. 77
(Eng. C.A.) ..................................................................................................... 163
Ron Brown Ltd. v. Johanson, [1990] B.C.J. 1923, unreported, August 10,
1990, Doc. No. Vernon 337/87 (S.C.) ............................................................... 86

## S

Scott Steel (Ottawa) Ltd. v. R.J. Nicol Construction (1975) Ltd. (1993), 15 C.L.R. (2d) 10 (Ont. Gen. Div.) .................................................. 91, 94, 96, 266, 267
Smith Bros. & Wilson (B.C.) Ltd. v. B.C. Hydro (1997), 30 B.C.L.R. (3d) 334, 33 C.L.R. (2d) 64 (S.C.) .................................................. 223, 225, 227, 233, 247
Sound Contracting Ltd. v. Nanaimo (City) (1997), 42 M.P.L.R. (2d) 202 (B.C.S.C.), revd (2000), 11 M.P.L.R. (3d) 49, [2000] B.C.J. No. 992 (C.A.), leave to appeal to S.C.C. dismissed without reasons [2000] S.C.C.A. No. 392 .................................................. 49, 245, 258, 278
Spectra Architectural Group Ltd. v. Eldred Sollows Consulting Ltd. (1991), 80 Alta. L.R. (2d) 361, 7 C.C.L.T. (2d) 169 (Alta. Master) .................................................. 192, 201
St. Lawrence Cement Inc. v. Farry Grading & Excavating Limited (1989), 32 C.L.R. 185 (Ont. H.C.), vard, [1992] O.J. No. 2220 .................................................. 159, 214
Stuart Olson Constructors v. NAP Building Products (1997), 33 C.L.R. (2d) 6 (B.C.S.C.) .................................................. 98

## T

Tarmac Canada Inc. v. Hamilton-Wentworth (Regional Municipality) (*sub nom.* George Wimpey Canada Ltd. v. Hamilton-Wentworth (Regional Municipality)) (1997), 34 C.L.R. (2d) 123 (Ont. Gen. Div.), affd (1999), unreported, September 13, 1999, Doc. No. C28200, McMurtry C.J.O., O'Connor, Sharpe JJA. (Ont. C.A.) .................................................. 44, 46, 237, 238, 239, 246, 258
Thomas C. Assaly Corporation Ltd. v. R. (1990), 44 Admin L.R. 89, 34 F.T.R. 156 (T.D.) .................................................. 209, 213, 216, 221, 244, 286
Thompson Bros. (Const.) Ltd. v. Wetaskiwin (City) (1997), [1998] 1 W.W.R. 101, 34 C.L.R. (2d) 197 (Alta. Q.B.) .................................................. 39, 238, 239, 277, 278
Toronto Transit Commission v. Gottardo Construction Ltd., Ontario Superior Court of Justice, December 2003 .................................................. 146, 292, 293
Town of Vaughan v. Alta Surety Company (1990), 42 C.L.R. 305 (Ont. H.C.), vard (1992), 3 C.L.R. (2d) 209 (Ont. C.A.), leave to appeal refd (1993), 147 N.R. 396n, 60 O.A.C. 217n (S.C.C.) .................................................. 119, 136
Twin City Mechanical v. Bradsil (1967) Limited (1996), 31 C.L.R. (2d) 210 (Ont. Gen. Div.), revd (1999), 116 O.A.C. 396, 43 C.L.R. (2d) 275 (C.A.), leave to appeal to S.C.C. dismissed without reasons [1999] S.C.C.A. No. 184 .................................................. 173, 175, 176, 179, 275, 283, 284

## U

United Services Funds v. Richardson Greenshields of Canada Limited (1988), 48 D.L.R. (4th) 98, 22 B.C.L.R. (2d) 322 (S.C.) .................................................. 170

## V

Vachon Construction Ltd. v. Cariboo (Regional District) (1994), 18 C.L.R. (2d) 109 (B.C.S.C.), revd (1996), 28 C.L.R. (2d) 145, 136 D.L.R. (4th) 307 (B.C.C.A.) .................................................. 123, 140, 224, 242, 252, 253, 261

Vic Van Isle Construction Ltd. v. School District No. 23 (1995), 25 C.L.R. (2d) 304 (B.C.S.C.), affd (1997), 144 W.A.C. 161, 33 C.L.R. (2d) 75 (B.C.C.A.), leave to appeal refd (1997), 163 W.A.C. 80n (S.C.C.) ......................... 127

Vipond Automatic Sprinkler Co. Ltd. v. E.S. Fox Ltd. (1996), 27 C.L.R. (2d) 311 (Ont. Gen. Div.) ............................................................................................. 95

## W

WIB Co. Construction Ltd. v. School District No. 23 (Central Okanagan) (1998), 41 C.L.R. (2d) 93, 44 B.L.R. (2d) 276 (B.C.S.C.), additional reasons, [1998] B.C.J. No. 2966, unreported, November 30, 1998, Doc. No. Kelowna 26578 (S.C.) .................................................................................. 198

Walford v. Miles, [1992] 2 A.C. 128, [1992] 1 All E.R. 453 (H.L.(E.)) ......................... 73

Walter Cabott Construction Ltd. v. R. (1974), 44 D.L.R. (3d) 82 (F.T.D.), vard (1975), 69 D.L.R. (3d) 542 (F.C.A.) ............................................................................. 158

Western Plumbing and Heating Ltd. v. Industrial Boiler-Tech Inc., [1999] N.S.J. No. 315, 48 C.L.R. (2d) 82 (S.C.) ..................................................................... 111

Westgate Mechanical Contractors Ltd. v. PCL Construction Ltd. (1987), 25 C.L.R. 96 (B.C.S.C.), affd (1989), 33 C.L.R. 265 (B.C.C.A.) ............................ 84, 97

Wigmar Construction (B.C.) Ltd. v. Defence Construction (1951) Limited (1997), 35 C.L.R. (2d) 185 (B.C.S.C.) ............................................................. 193, 195

Wind Power Inc. v. Saskatchewan Power Corp., [2002] S.J. No. 287, 2002 SKCA 61, [2002] 7 W.W.R. 73, 217 Sask. R. 193, 24 B.L.R. (3d) 197, 15 C.L.R. (3d) 291, leave to appeal to S.C.C. dismissed without reasons [2002] S.C.C.A. No. 283 ............................................................................................ 68

# Chapter 1

# TRADITIONAL LAW OF BIDDING

*Before we talk about the law as it relates to competitive bidding, let us have a quick look at some of the basic legal principles as they apply to contracts. A case that has been in legal textbooks for over a hundred years will serve as an illustration.*

## DICKINSON v. DODDS

House of Lords; 1874

In 1874, George Dickinson was considering buying a house belonging to John Dodds but could not quite make up his mind. So, on June 10, Dodds wrote a letter to Dickinson. Its key part read as follows:

> I hereby agree to sell to Mr. George Dickinson the whole of the dwelling-houses, garden ground, stabling, and outbuildings thereto belonging, situate at Croft, belonging to me, for the sum of £800. As witness my hand this tenth day of June, 1874.
>
> (Signed) John Dodds
>
> P. S. This offer to be left open until Friday, 9 o'clock, A. M., 12th June, 1874.

The day the letter was signed and delivered was a Wednesday. The following day, Dickinson was informed that Dodds had offered to sell the property to a Thomas Allan. This prompted Dickinson to deliver immediately to Dodds' mother-in-law a formal acceptance of the offer to sell the property. However, she forgot to give the document to Dodds.

The document reached Dodds early on Friday morning, but Dodds rejected it, saying that it was too late. He had already sold the property to Allan. Dickinson sued for breach of contract, and the whole matter ended up in front of the highest court in England, the House of Lords.

Was there a contract between Dickinson and Dodds? Two people enter into a contract when one of them makes an *offer* and the other *accepts* it. The document, though beginning with "I hereby agree to sell", was nothing but an offer, for Dickinson himself required time to consider whether he would enter into an agreement or not. Lord James decided that since one of the parties had not agreed to the sale at the time Dodds wrote his letter, there was no contract.

What of the promise to keep the offer open? If a contract is to be legally enforceable, the person accepting an offer must give something in exchange (or the contract must be under seal). That "something" is called *consideration*. There is a Latin name for a contract without consideration: *nudum pactum*. Such a "nude" contract is unenforceable.

Dodds received no consideration for his promise to keep the property unsold until 9 a.m. on Friday. Said Lord James:

> It is clear settled law, on one of the clearest principles of law, that this promise, being a mere nudum pactum, was not binding, and that at any moment before a complete acceptance by Dickinson of the offer, Dodds was as free as Dickinson himself.

Lord Mellish concurred in the decision. In order to make a contract, the two parties must be in agreement at the same time. Dickinson knew that Dodds had sold the property to someone else; therefore, he knew that Dodds had changed his mind. How could he then accept Dodds' offer and thereby make a valid contract? "It seems to me that would be simply absurd", said His Lordship.

―――――――

*Until 1981, the decision of the House of Lords in this classic case could be applied to competitive bidding in Canada.*

*Traditionally, the call for bids was an invitation to potential bidders to submit offers. The bids were considered to be **offers**, and the **acceptance** of a bid by the owner created a **contract** between owner and bidder.*

*A bidder's offer, clearly, could be withdrawn at any time before the owner accepted it. Even if the bidder promised that it would not revoke its bid, it couldn't be held to this promise in law because there had been no **consideration** for the promise — the bidder got nothing in return for its promise.*

*It follows also that an owner could not accept a bid that it knew contained a fundamental mistake. Furthermore, the owner was not bound to accept the lowest (or any) bid, regardless of whether this was stated in the bid call or advertisement.*

*"Suppose there had been only one tender, would the [owner] have been bound to accept that? The advertisement clearly does not amount to a contract; it only invites offers", decided the court in an even older case: Spencer v. Harding, of 1870 vintage.*

*Let us now have a look at how these common law principles were applied in a couple of cases where the contractor had made a significant mistake in its bid.*

# MCMASTER UNIVERSITY v. WILCHAR CONSTRUCTION LTD.

Ontario High Court of Justice; August 1971

In 1969, McMaster University invited bids for the construction of a health sciences centre in Hamilton, Ontario. Wilchar Construction Limited, one of five contractors, took up the invitation and presented its bid a few minutes before closing time.

All bids were made subject to a wage escalator clause. This was important because, at the time, certain unionized workers were out on strike for higher wages. The Hamilton Construction Association suggested the use of the wage escalator clause which would allow for payment of union rates as eventually settled between the striking unions and the employers.

In preparing its bid, Wilchar planned to insert the escalator clause on the first page as required. However, due to some last-minute confusion in the office, the first page was omitted from the bid, and the bid was submitted without the clause.

The bid form prepared by McMaster University consisted of only nine pages. At the trial, Justice Thompson decided that it must have been apparent to the owner that the first page was missing from the Wilchar bid.

Of the total bid sum of $185,613, the contractor allowed only the sum of $10,500 for profit and overhead. Evidence later indicated that this was much less than the increased cost of labour under the escalator clause. Without that clause, therefore, the contractor would have lost a substantial sum of money.

Shortly after the opening of the bids, Wilchar discovered its error and informed the owner. The information was not received in a helpful spirit; the owner was not prepared to let the contractor add the escalator clause. Justice Thompson commented:

> I have been unable to escape the conclusion that [the representatives of the owner] were intent at all costs upon acquiring some easy money for the [owner] at Wilchar's expense.

The owner insisted on accepting Wilchar's bid as it was while, at the same time, requiring the contractor to ensure that all workers be paid at the prevailing union rates. Understandably, Wilchar refused to sign the contract when the owner demanded that it do so.

At trial, the judge made some scathing comments regarding the owner's evidence. There was no doubt in the judge's mind that the real reason the owner purported to accept Wilchar's bid was the hope that it might be able to grab the penalty of the bid bond. It knew full well that Wilchar had made a mistake in its bid and that the contractor would refuse to enter into a contract unless the error was corrected.

Counsel for the owner conceded that Wilchar had made a mistake but argued that this was not a mistake of a fundamental character; it was merely a mistake in the motive or the reason for making the bid.

The court did not accept the owner's argument. In a construction contract, the price is *always* a fundamental term of the contract, said the judge. In fact, it is the very basis of such a contract. If the mistake affects a fundamental term of the contract and the other party knows about it, the contract is void in law.

There could be little doubt that there was no real agreement between the parties. One party had intended to make a contract on one set of terms, and the other had intended to make it upon another set of terms with the result that there was a lack of consensus. Therefore, there was no contract.

The court decided that, since there was no legally enforceable contract between the owner and the contractor, Wilchar was under no obligation to perform it.

The case was appealed, but the Court of Appeal affirmed the decision.

---

*In the case of a mistake "on the face of the bid" — one that is readily seen by the person receiving the bid — the decision in Wilchar is still the law, in spite of the changes that have happened since then.*

*The other often-quoted case involving a mistake in the contractor's bid is also one of the last cases to be based on the traditional concept of the bidder being the **offeror** and the owner being the **offeree** — before the 1981 revolution in the law of bidding changed everything and, incidentally, made this book necessary.*

*The case in question follows below.*

## BELLE RIVER COMMUNITY ARENA INC. v. W.J.C. KAUFMANN CO. LTD.

Ontario Court of Appeal; June 1978

In 1972, the Town of Belle River in Ontario sent out invitations to bid on the proposed construction of a community arena. Bidders were required to furnish a bid bond of $40,000 or a certified cheque in that amount. The bids were to remain open for acceptance for a period of 60 days.

W.J.C. Kaufmann Co. Ltd. put in a bid for the job, accompanied by a bid bond issued by the United States Fidelity and Guaranty Company.

Bids were opened on January 11, 1973. Kaufmann's bid was the lowest of nine bids by just over $15,000. The owner did not accept the bid that day. The following morning, Kaufmann sent a telegram to the architects informing them that it wished to withdraw the bid because it contained a serious error: the person preparing the bid had left out a summary sheet.

Nevertheless, more than a month later, the owner instructed its solicitors to advise Kaufmann that its bid had been accepted. The solicitors for Kaufmann reiterated their position that their client had made a mistake and had withdrawn the bid before acceptance. The owner never submitted Kaufmann a formal construction contract but entered into a contract with the next lowest bidder and sued Kaufmann for the difference between the two bids.

The trial judge dismissed the owner's claim because Kaufmann had not formally refused to enter into a contract before the expiration of the period during which all bids were open for acceptance.

The judge also commented on the defence that the owner could not have effectively accepted the Kaufmann bid, even though it was irrevocable for 60 days, because the owner knew that the contract price in the bid was substantially lower than intended because of a mistake. The judge quoted from *Corbin on Contracts*:

> Courts refusing to decree rescission for unilateral mistake often say that to do otherwise would tend greatly to destroy stability and certainty in the making of contracts. In some degree, this may be true; but certainty in the law is largely an illusion at best, and altogether too high a price may be paid in the effort to attain it. Inflexible and mechanical rules lead to their own avoidance by fiction and camouflage.
>
> A sufficient degree of stability and certainty will be maintained if the court carefully weighs the combination of factors in each case, is convinced that the substantial mistake asserted was in fact made, and gives due weight to material changes of position. Proof of the mistake should be required to be strong and convincing, but in many of the cases it is evident that such proof existed.

In the Kaufmann case, the existence of the mistake was not questioned. There was no time for the owner to do anything before it was informed of the mistake.

> **Rescission**. The setting aside or cancellation of a contract. The contract which is *rescinded* is treated as if it had never existed.

Belle River appealed the decision, but the Ontario Court of Appeal dismissed the appeal.

"In my view", said Justice Arnup of the Court of Appeal, "the authorities establish that an offeree cannot accept an offer which he knows has been made by mistake and which affects a fundamental term of the contract. ... In substance, the purported offer, because of the mistake, is not the offer the offeror intended to make, and the offeree knows that".

The principle applies even if there is a provision binding the offeror to keep the offer open for acceptance for a given period. The situation is different when the offeree does not know the offer is made by mistake and accepts the offer, taking what it plainly says as its face value.

In this case, the court decided the owner knew right from the day after the bids were opened that a serious and material error as to price had been made in Kaufmann's bid. "From that time on, the plaintiff could not accept that bid, as a matter of law", said Justice Arnup.

Counsel for the owner argued that the bond, in plain language, made both Kaufmann and the bonding company liable because, unless all of the conditions of the bond were satisfied, the obligation to pay remained. The court interpreted the effect of the bond differently. If the contractor's bid is accepted, and the contractor refuses to enter into a formal contract to do the work, the contractor and the bonding company are obliged to pay the owner any extra amount up to the maximum amount of the bid bond that the owner must pay to have the work done.

It is true that the bidders had contracted to keep the bid open for acceptance for the full term of 60 days. But it did not matter whether the bid could be withdrawn or was, in fact, withdrawn:

- If Kaufmann's bid could be withdrawn before acceptance, then it was withdrawn, and no contract came into existence.
- If it could not be withdrawn for 60 days, it, nevertheless, could not be accepted because the owner knew of the mistake.

Therefore, no contract came into existence, and, under the bid bond, the contractor was not liable to the owner.

## SUMMARY

The two cases we have seen so far make sense and conform to traditional principles of common law. To summarize:

- if A offers to sell B a car for $2,000 and B says "I accept", there is a *contract*;
- there is no enforceable contract without *consideration*: A's consideration is the money, B's consideration, the car;
- A can withdraw its offer at any time before B accepts it;
- if B says "I accept your offer, but will only pay $1,800 for the car", there is no contract — B has made a *counteroffer* and the bargaining process starts from scratch;

- the *counteroffer* is, by operation of law, a rejection of A's offer — it is no longer capable of acceptance;
- if A's offer contains a mistake that affects a *fundamental term* such as price, and if that mistake is known to B, B cannot "snap up the offer" and form a valid contract — where there is no agreement as to terms, there can be no contract; and
- calling for bids used to be considered an invitation to submit offers — a contract would be formed when the owner accepted a bid; then all the principles of contract law would apply. This used to be so, and this still is the law in all other common-law countries — but not in Canada.

One of the basic principles of common law — that the maker of an offer may withdraw its offer anytime prior to acceptance — created a serious commercial problem for building owners seeking irrevocable bids for their projects. As we shall see in the next chapter, the Supreme Court of Canada solved the industry's problem, no doubt about that. But, the remedy the Court created produced side effects, and is still producing them today, and fundamentally changed the face of bidding in Canada.

# Chapter 2

# REVOLUTION IN THE LAW OF BIDDING

The bidders' freedom to withdraw their bids up to the moment the owner accepted one of them sometimes created problems. The lowest bidder, if it thought its bid too low, would simply withdraw without suffering any legal sanctions. An unscrupulous contractor, knowing that it was free to withdraw its bid at any time before the bid was accepted, could submit a "guesstimate" for a project instead of a serious bid.

The problem was even worse for general contractors that relied on subcontractors' quotations to compile their bids to the owner, and for subcontractors that relied on sub-subcontractors, and so on, down the construction pyramid.

The Belle River case is a good illustration of the tension that existed in construction bidding. On either analysis in the case, the owner lost. It seemed a bit of a farce to watch a contractor submit an irrevocable bid with bid security and walk away unbound and unscathed by simply crying "mistake". Everybody recognized that there was something wrong with this picture.

The construction industry never managed to craft a solution to this dilemma. The Supreme Court of Canada finally resolved it in its own way, and the industry has had to live with that solution, whether it liked it or not.

A form of earthquake hit in 1981, in a case involving a mistaken bid. Since then, virtually no discussion of competitive bidding can take place without the name of *Ron Engineering* being mentioned.

This case represents a major upheaval in the law as it applies to bidding and tendering. Well over a century of legal precedents and tradition have been turned upside down. The answers to the many questions raised by this revolution are still coming in; quite a few are supplied by the cases summarized in this book.

Some important questions are likely to remain open for a long time yet. When the court in *Ron Engineering* used the so-called Contract A to reach its decision, it spawned a cottage industry whose mission is to define and expand the boundaries of Contract A.

## R. v. RON ENGINEERING & CONSTRUCTION (EASTERN) LTD.

Supreme Court of Canada; January 1981

The contractor, Ron Engineering & Construction (Eastern) Ltd., submitted a bid along with a cheque for $150,000 to the Ontario government for a construction project. This submission represented a bid deposit required by the Instructions to Bidders.

The contractor's employee who filed the bid remained for the opening of the bids and discovered that Ron's bid (amounting to $2,748,000) was the lowest and that the difference between it and the next lowest bid was over $600,000. The employee immediately informed the contractor, who uncovered a serious error in the bid. Just over an hour after the opening of the bids, the contractor sent a telex to the owner, notifying it of the error and requesting to withdraw the bid without penalty.

The owner, nevertheless, awarded the contractor the construction contract. The contractor refused to sign. Its position was that since it had notified the owner of the error prior to accepting the bid, the bid could not be accepted as the basis of a contract.

The owner, however, decided that the contractor's refusal entitled the owner to retain the bid deposit. The owner then accepted the second lowest bid. The contractor sued to recover the money. The owner counterclaimed for damages caused by its having to accept a higher bid.

The trial judge found that the owner was entitled to retain the bid deposit, and the judge, therefore, dismissed the counterclaim. The contractor appealed the decision. The Court of Appeal concluded that the owner could not have accepted an offer that it knew had been made by mistake, when the mistake affected a fundamental term of the contract. The judge stated that an owner calling for bids:

> ... is entitled to be skeptical when a bidder who is the low tenderer by a very substantial amount attempts to say, after the opening of tenders, that a mistake has been made. However, when that mistake is proven by the production of reasonable evidence, the person to whom the tender is made is not in a position to accept the tender or to seek to forfeit the bid deposit.

The owner then appealed to the Supreme Court of Canada. Justice Estey, who wrote the decision for the court, found that the process of bidding involves *two contracts*, which he called Contract A and Contract B.

*Contract A* (or *bidding contract*) arises between the contractor and the owner immediately when the bid is submitted, and without further formalities. This is a so-called *unilateral* contract, said the judge, which comes into being when the contractor submits a bid. If several contractors bid on the project, the owner automatically enters into separate contracts with each one of them.

A unilateral contract, said Justice Estey, results from an act made in response to an offer as, for example, "I will pay you a dollar if you will cut my lawn". There is no obligation on anyone to cut the lawn, but if someone does actually cut the lawn, the offeror is obliged to pay up. The owner's call for bids is such a unilateral offer, and the submission of a bid in response to the call represents acceptance of the offer.

The terms of this "bidding contract" are contained in the bid documents issued with the call for bids. The principle is the irrevocability of the bid, if that is what the bid documents require. From the moment the bids are submitted, the owner and each of the bidders are locked in a bidding contract. According to Justice Estey:

> The significance of the bid in law is that it at once becomes irrevocable if filed in conformity with the terms and conditions under which the call for tenders was made and if such terms so provide.

So, a bid submitted under these circumstances is no longer a "nudum pactum" — as it used to be. Instead, the bid now exists within a contractual envelope — Contract A — which provides that the bid will not be withdrawn during the period of irrevocability.

A corollary term is the obligation imposed on both parties to enter, if the bid is accepted, into a *Contract B* (the *construction contract*).

So, under Contract A, Ron Engineering could not withdraw its bid for the period specified in the bid documents. The bid deposit, in the view of the court, was there to ensure that the bidder performed its obligations under Contract A, which was the only contract in existence at that time.

Neither Ron Engineering nor the owner was aware of any mistake up to the moment the bid was submitted and Contract A came into existence. The contractor intended to — and did — submit the bid as it was, including the price that was stipulated in it.

There was nothing on the face of the bid to reveal that there was an error. The amount of the bid did not reveal a major discrepancy because a cost estimate, prepared for the owner by a firm of engineers, was, in fact, slightly lower than Ron Engineering's bid containing the error.

In this respect, the case differed essentially from *Wilchar Construction.* In that case, the bidder mistakenly omitted one page of the bid. The judge found that it must have been apparent to the owner that this page was missing from the bid. The judge commented:

To me, this is patently a case where the offeree, for its own advantage, snapped at the offeror's offer well knowing that the offer as made was made by mistake.

The court, therefore, dismissed McMaster's claim for damages (equal to the difference between the defendant's bid and the next lowest), finding that the bidder was not obliged to conclude a construction contract. The important fact in that case was that the document submitted by the contractor was, on its face, incomplete and could not in law amount to a bid as required by the conditions established in the call for bids.

No such obvious mistake existed in the bid submitted by Ron Engineering. The effect of the mistake in the contractor's calculations on Contract B was an entirely different question, and the Supreme Court felt no need to resolve it. The appeal was concerned only with the claim made by the contractor for the return of the bid deposit. The terms of Contract A clearly indicated that the owner had a contractual right to the deposit.

---

*As we saw in the very first case in this book, Dickinson v. Dodds, there are three basic ingredients to a contract: offer, acceptance and consideration. For example, I offer to sell you my car for $10,000. If you accept, we have a contract. You get the car, that is your consideration; I get the money, that is mine.*

*The bidding process is far less clear cut.*

*The owner kicks off by inviting bids to construct the project. The bidders submit their offers, and when the owner picks and accepts one of those offers, a construction contract is formed. The consideration is evident: the contractor gets its money, the owner gets a completed project. What emerged for the first time in Ron Engineering is that the construction contract is only the second layer in a two-layer process: this is* **Contract B**.

*Contract B has only to do with* **what** *is going to be constructed. The plans and specifications in the bid documents define that.*

*The first layer, which Justice Estey called Contract A, has only to do with* **how** *the bidding process itself is conducted.*

*The owner's invitation to bidders does more than just ask for offers. It is itself an offer. The owner's offer is to step out onto the dance floor — notionally at least — with each of the bidders. The dance is the bid process. When a bidder accepts the owner's offer and agrees to dance (by submitting a compliant bid),* **Contract A** *is created.*

*This is not some casual encounter — after all it is a contract. So, this dance has rules. The rules — the dance steps — are set out in the bid documents, usually in the Instructions to Bidders and the terms of Contract B offered by the owner. Contract A has nothing to do with the project or its construction. Contract A describes the dance — and the steps to be followed.*

*By the time the music stops, the owner has to make a choice. It is free to say good evening to all of its dance partners and go home alone. Or, it may select one partner with whom it wants a permanent relationship — in the form of Contract B. As will be pointed out later in this book, the owner is not free (usually) to abandon all of its partners and take up (award Contract B) with someone who was not invited to the dance in the first place — or refused to come.*

*On the flip side, none of the owner's dancing partners is free to leave the dance or to reject the invitation of the owner to continue their encounter. That right was given up when they agreed to dance in the first place.*

*But back to the construction setting. Let us try and paraphrase the owner's invitation to bidders in terms of the Contract A offer, acceptance and consideration.*

*Here is my offer: I offer you a chance to compete for my project. To accept my offer, submit a package which (more or less precisely) follows the rules and requirements set out in my offer. If you do that, we have a contract. This is our Contract A.*

*Note that both parties receive consideration: the bidder gets a chance to compete, and the owner gets the package according to the rules. What is in the package is immaterial at the Contract A stage. Contract A is concerned only with the packaging, delivery and processing, not the content. That is why, if the bidder has made a mistake in its bid, the bidder has to live with it; there is no mechanism under Contract A to correct it (unless the bid is an obvious dud).*

*Bids that don't follow the rules of the offer cannot be the basis for a Contract A because the bidder has not accepted what the owner offered — offer and acceptance must (more or less precisely) match, or there is no contract.*

*What's in the bid? It contains the bidder's offer for the project. This offer, if accepted by the owner, will be the basis of a new contract — Contract B.*

*When the owner opens the bids, checks them for compliance with the rules, selects the best compliant offer and awards the construction contract, the owner is bound all along by the terms of Contract A (just as the bidder is bound).*

*Finally, both the owner and the successful bidder sign on the dotted line of the construction contract. The bidder's offer has been accepted, Contract B is now in place and Contract A is at an end. The consideration for Contract B is obvious: the bidder will do the required work, and the owner will pay the contract price — too bad Contract A is not that simple.*

*The importance of the Ron Engineering decision, however, did not sink in right away. It took the construction industry a while to realize that the Contract A/Contract B scenario was there to stay. The following decision discussed here made that clear. It involved another contractor that had made a mistake and had the mistaken bid accepted by the owner.*

## CALGARY (CITY) v. NORTHERN CONSTRUCTION COMPANY LTD.

Alberta Court of Appeal; December 1985

In 1978, the City of Calgary invited bids for a construction project. The bid documents stipulated that once all bids had been opened, they would be irrevocable until the successful bidder signed the construction contract.

Northern Construction submitted a bid accompanied by a bid bond and a consent of a surety to furnish a performance bond. When the bid documents were opened, Northern, with a bid of $9,342,000, was the lowest of nine lump-sum bids. The next lowest bid was $395,000 more.

When the bids were opened, Northern had an employee present who reported the outcome to the contractor's office. Shortly before 4:00 p.m. on the same day, two hours after the bid had been opened, a clerical error of over $181,000 was discovered. Thereafter, Northern immediately notified the City of Calgary.

The error had not been apparent "on the face of the bid".

At a meeting held with the representatives of the City the following day, Northern requested that the amount of $181,274 be added to its contract or, alternatively, that Northern be allowed to withdraw its bid and have its bid bond returned. The same request was later made by letter. Northern stressed the fact that if an amendment of the bid were accepted, its bid would *still be less* than the next lowest bid.

Calgary refused the request and accepted Northern's bid. When Northern declined to execute the construction contract, the City awarded it to the second lowest bidder, Pigott Construction, and sued Northern for damages.

There were many similarities between the *Ron Engineering* case and the *Northern Construction* case. However, the trial judge found dissimilarities that, in his opinion, rendered the precedent inapplicable. The crucial point was that Calgary had accepted the contractor's bid. That acceptance removed the case from a Contract A (bidding contract) situation and created a Contract B (construction contract) situation with which the *Ron Engineering* case was not concerned.

Consequently, the trial judge decided the case by applying the traditional law of mistake. At the time Calgary purported to accept the bid of Northern Construction, the City was fully informed that the contractor had made a fundamental mistake in its bid. Therefore, decided the judge, Calgary could not legally accept the bid.

The Alberta Court of Appeal took a different view of the case. The majority of the court held that Contract B had never come into existence because it had not been executed by both parties.

The correct interpretation of the situation was that the City, in insisting on the signing of the construction contract, was exercising a right granted to it by Contract A, said Justice McDermid, who added:

> If this is the correct view, the contractor is placed in a dilemma, for if he executes Contract B I do not think he could then raise the question of mistake, while if he does not do so, he is in breach of Contract A.

In other words, damned if he does, damned if he doesn't!

The contractor tried another approach to defend its case. Northern argued that the City had been under a duty to *mitigate its damages*, therefore, it should have accepted its offer to carry out the construction for its original offer plus the amount of the error, because that would still have been less than the next lowest bid.

---

**Mitigation of Damages.** The party that seeks to recover damages for breach of contract or negligence must take all reasonable steps to minimize the loss or injury it may suffer. This is consistent with the general rule that a person cannot recover damages that could reasonably have been avoided.

---

True, said the Court of Appeal, the City did have the duty to mitigate. However, to accept the argument of the contractor would be to change the bidding system to an auction. It would allow any contractor that made a low bid to refuse the contract but to offer to do the work for less than the second lowest bidder and then to argue that the owner must accept such offer in mitigating its damages. "The city was under no such duty and the contractor has not proven any failure of the city to mitigate", decided Justice McDermid.

The amount of $395,000 claimed by Calgary as damages, namely the difference between Northern's bid and the next lowest, was the actual damages proven by the City. The majority of the court, therefore, allowed the claim.

Quite apart from the rules regarding the formation of contracts, would it be unconscionable in *equity* for the City to accept Contract B after discovering that the contractor's bid was the result of an innocent and honest error of fact?

Justice Kerans of the Court of Appeal found the preponderance of legal authority supported the view that equity would not intervene to protect a party from a bad bargain:

> Where a claim is made that a bargain is unconscionable, it must be shown for success that there was inequality in the position of the parties due to the ignorance, need or distress of the weaker, which would leave him in the power of the stronger, coupled with proof of substantial unfairness in the bargain. When this has been shown a presumption of fraud is raised and the stronger [party] must show, in order to preserve his bargain, that it was fair and reasonable.

**16** BIDDING AND TENDERING: WHAT IS THE LAW?

In the dispute between Northern and Calgary, the bargaining power of the two parties was not so disproportionate as to put Calgary in a situation where its claim would be unconscionable.

Northern appealed the decision to the Supreme Court of Canada, and so the top court had an opportunity to reconsider *Ron Engineering*. It did not reconsider it. In 1987, the Supreme Court, in a very brief decision, unanimously affirmed the decision of the Alberta Court of Appeal and found that the facts of the case could not be distinguished from *Ron Engineering*; the case was governed by that precedent.

---

*The revolutionary rule in the law of bidding and tendering was now fully in command. All that remained was to apply the bare-bones principles of Ron Engineering to the hundreds of practical problems that happen in connection with bidding and tendering in construction, especially in the hundreds of legal actions across Canada. The most important of these actions that made it to trial and were decided by the courts are summarized here.*

*Is Ron still with us?*

*Only a few years after Ron Engineering, the Supreme Court of Canada had its first opportunity, as mentioned earlier, to review this drastic decision in the Northern Construction case discussed above. It affirmed the Ron Engineering reasoning.*

*In 1999, the top court once again had an important bidding case before it: M.J.B. Enterprises v. Defence Construction, which is discussed later in this book. Those hoping for the demise of the Contract A/Contract B paradigm were disappointed, but not entirely. Ron was bent but not broken. The Contract A concept survived, although its existence is not a given.*

*Justice Iacobucci, speaking for the Supreme Court of Canada, stated (paragraph 19, emphasis added):*

> What is important... is that the submission of a tender in response to an invitation to tender **may give rise** to contractual obligations, quite apart from the obligations associated with the construction contract to be entered into upon the acceptance depending on whether the parties intend to initiate contractual relations by the submission of a bid. If such a contract arises, its terms are governed by the terms and conditions of the tender call.

*M.J.B. was a "garden variety" bid case in which the Supreme Court of Canada found Contract A did arise.*

*One does not have to be clairvoyant to anticipate future bid documents which stipulate that the submission of a bid does not give rise to Contract A or does not impose any kind of obligation on the party calling the bid. Whether such a disclaimer will work depends on the circumstances. But, for the "garden variety" of bid call, Contract A lives.*

In 2000, the Supreme Court of Canada ruled on Martel Building Ltd. v. Canada. Confirming the view of numerous lower courts, the Supreme Court of Canada stated that an implied duty of fairness is a term of Contract A unless it is expressly excluded. (See the Martel decision in Chapter 3.)

In 2001, the same Court was presented with a bidding case involving a contractor which had carried a particular trade in its bid to the owner. (See the Naylor Group Inc. v. Ellis-Don Construction Ltd. decision in Chapter 4.) Again, confirming the view held by numerous lower courts, the Supreme Court of Canada extended the Contract A/Contract B paradigm into contractor/subcontractor bid dynamics.

What is interesting about the history of Ron Engineering — and its successful run in the Supreme Court of Canada — is that the last three appearances (M.J.B., Martel and Naylor) have had nothing to do with contractor error. The law around mistake seems well settled — the outcome in each case being determined by the facts.

As will be seen in what follows in this book, contractor error was quickly bumped out of the spotlight once the implications of the Ron Engineering decision became better understood.

## SUMMARY

The *Ron Engineering* revolution can be summarized as follows:

- The Supreme Court of Canada in its *Ron Engineering* decision split the process of bidding into two distinct contracts: Contract A and Contract B.
- When the owner invites bids, it offers to enter into a "bidding contract" (Contract A) with each and every bidder that submits a bid which follows the rules in the bid documents.
- The consideration the contractor receives is the owner's implied promise that if the bidder follows the rules set out in the bid documents, the bid will be evaluated for the construction contract.
- The bidding contract (Contract A) is formed when each bidder submits its bid on time and in the proper form, thus accepting the owner's offer.
- The owner has a binding and legally enforceable contract with each compliant bidder, and the owner can be liable for damages if it breaches this contract.
- If the bid documents call for irrevocable bids, then the bidder, by submitting its bid, accepts this term of Contract A and agrees not to alter or withdraw its

bid until a specified time has elapsed even if the bidder realizes after closing that it has made a mistake.
- The bidder agrees to enter into a formal construction contract (Contract B) if the owner awards it to him.
- If the bidder breaches any of these terms of Contract A, it can be liable in damages.
- The submission of a compliant bid represents both acceptance of the owner's offer to enter Contract A, and an offer to perform the work stipulated in the bid documents.
- The owner's acceptance of the bidder's offer under the rules of Contract A puts the construction contract (Contract B) in place.

# Chapter 3

# DUTIES AND RIGHTS OF OWNERS

*So far, we have seen the effect of Contract A on the duties and obligations of the bidders. The duties and obligations of owners are just as firmly fixed by the bidding contract as those of contractors. The Ron Engineering decision did not have much to say about these, but the picture became clearer in the decisions that followed.*

*You will notice that all of the owner/contractor bid litigation cases summarized below involve an owner that is a public body ranging from a not-for-profit corporation to a community college to one of the three levels of government. Within this spectrum, the levels of government have occupied most of the starring roles.*

*This is not the result of a deliberate selection. These are the leading cases that fall into this chapter.*

*But, back to the bidding contract. When courts are faced with circumstances that the contracting parties have not specifically addressed, the courts often trot out the "officious bystander" to help them out. The courts imagine that the bystander asks the parties, at the beginning of their relationship, what the parties would do if a situation which is not covered by their contract arose. The courts then imagine how the parties would answer that question. The answer is usually treated as though it was in the contract (an implied term).*

*If the people in whose name the government acts were asked by the officious bystander what standard government representatives should bring to the purchase of construction services, the answer would undoubtedly be: "Obtain the best price possible for us within the bounds of fairness."*

*Some people demand that the government lead by example. In the bidding process, government owners certainly have led by example. Based on what the case law tells you, is it a good example?*

## BEST CLEANERS AND CONTRACTORS LTD. v. R.

Federal Court of Appeal; March 1985

In 1981, the federal Department of Transport published an invitation to bid for the operation and maintenance of the airport at Frobisher Bay, N.W.T. The bid documents offered a contract for a two-year period. In addition, bidders were requested to provide an indicative price — not a firm offer — for a further period of two years.

Only two bids were accepted by the department. One was from Best Cleaners and Contractors Ltd. and the other, from Tower Arctic Limited. Best Cleaners' bid for the initial period was $948,000, approximately $4,500 less than Tower's, but its price for the extended period exceeded Tower's by more than $60,000.

After the bids were opened, an official of the department contacted Tower and asked whether the company would agree to enter into a four-year contract as tendered. When Tower agreed, the department recommended to the Treasury Board that the contract be awarded to Tower for a period of four years. When Best Cleaners heard of that recommendation, it wrote to the department to complain that departmental officials had no right to negotiate with either of the bidders for a change in the terms of the proposed contract.

The Treasury Board sought legal advice and apparently discovered that to award the contract to Tower for a four-year period would be illegal. The Board still awarded the contract to Tower but only for a period of two years as provided in the bid documents. Best Cleaners sued, claiming that the award was made in bad faith because the decision was not based on the relative merits of the bids or the relative abilities of the bidders, but on "illegal" negotiations with Tower.

Following some legal manoeuvres of no relevance to the construction issues, the case ended up in the Federal Court of Appeal.

The relationship between the owner and the bidder for a building contract is governed by the 1981 decision *Ron Engineering*. When a contractor submits its bid in response to a call for bids, a bidding contract immediately arises between the owner and the bidder — the so-called Contract A. The owner's obligation under Contract A is not to award a construction contract (Contract B) except in accordance with the terms of the bid call.

The bid documents contained a clause saying that "the lowest or any tender need not be accepted" — the so-called privilege clause. The court held that the clause did not change the owner's obligation. The owner could award no contract at all, or it could award Contract B to Tower, but, under Contract A, it was under obligation to Best Cleaners not to award to Tower something *other* than Contract B.

The evidence clearly established that the Minister of Transport, on the advice of officials, recommended to the Treasury Board that it should breach Contract A by awarding Tower a four-year contract. A two-year contract was eventually signed, but the court decided that this contract was a sham — a two-year contract in form but a four-year contract in substance.

Justice Pratte found that Contract A included certain *implied terms*, and he made a statement destined to be repeated in many future decisions (emphasis added):

> In my view, [the implied terms of the bidding contract] simply impose on the owner calling the tenders the obligation to *treat all bidders fairly* and not to give any of them an unfair advantage over the others.

In negotiating with Tower and awarding it a contract that differed from what the bid documents called for, the minister breached this duty to treat all bidders fairly. The appeal by Best Cleaners was allowed.

*Thus, the owner's obligation under the bidding contract (Contract A) is not to award a construction contract (Contract B) except in accordance with the terms set out in the bid documents*

*So, the owner is not allowed to unilaterally change the bid documents before accepting the bid. Rather, the owner must accept or reject the bid as submitted, and, if necessary, make changes later in the construction contract.*

*Bidders are entitled to rely on the specifications in the bid call as remaining essentially unchanged. As in the Best Cleaners case, the overriding concern of the court is the need to treat all bidders fairly.*

*One of the thorniest issues in bidding is the effect of the so-called privilege clause that, in one form or another, can be found in most bid calls:* "**The lowest or any bid will not necessarily be accepted.**"

*Does this mean the owner can do what it likes with the bids? Or is the industry custom more important, which, except in limited circumstances, would award the contract to the lowest bid?*

*For a long time, the answer to the question depended on which province the court was located in. First we'll look at a solution "made in Ontario".*

## ELGIN CONSTRUCTION LTD. v. RUSSELL TOWNSHIP

Supreme Court of Ontario; January 1987

In 1983, Russell Township invited bids for the construction of water mains and sewers. The advertisement contained the so-called privilege clause:

> The Township reserves the right to reject any and all tenders, and the lowest or any tender will not necessarily be accepted.

The bid documents available to all bidders contained a similar warning.

The bid submitted by Elgin Construction was the lowest, but it provided for a completion time of 52 weeks. The second lowest bid was submitted by Atomik

Construction and provided for a completion time of only 28 weeks. This meant that if Atomik's bid was accepted, the total cost to Russell would be less because of the shorter period engineers would have to supervise.

The consulting engineers of the Township, with support from the township council, suggested to Elgin that it should qualify its bid by reducing the completion time to 28 weeks, without changing the price. Elgin complied, but the Township, nevertheless, accepted Atomik's bid. Elgin sued for damages.

The key ground on which Elgin pursued its action was that Russell, in rejecting Elgin's bid, failed to follow a "custom of the trade". Two professional engineers submitted affidavits in support of Elgin. According to the engineers, the contract is generally awarded to the lowest bidder unless unusual circumstances exist that would justify the refusal to do so.

The engineers testified that the clause in the invitation to bid that gives the owner the option to reject any or all bids is to protect the owner from having to award the contract to the lowest bidder when:

- all bids are too high;
- the lowest bidder lacks expertise; and
- some irregularity exists in the bidding process.

Justice White rejected this argument and said:

> It is my opinion that no "custom of the trade" can be deemed to qualify the most explicit words of the advertisement that ... "The Township reserves the right to reject any and all tenders, and the lowest or any tender will not necessarily be accepted," and the equally explicit words in the "Information for Tenderers."

These words form a legal context in which the contractor's bid was submitted. As a matter of law, said the judge, that context precludes the operability of the custom of the trade:

> To deny this would be to destroy the doctrine that contractual relations between parties are based on their objective manifestations of intent to exchange binding promises.

The court found that Elgin's action was "utterly without merit" and that there was no genuine issue for trial.

---

*A few months after the Elgin decision in Ontario, a County Court in British Columbia went its own way. The decision of Judge Selbie in the dispute between Chinook Aggregates Ltd. and the Municipal District of Abbotsford caused quite a stir because the learned judge preferred to follow the industry custom rather* than the letter of the contract. The British Columbia Court of Appeal affirmed the decision of Judge Selbie, and its reasoning is summarized below.

## CHINOOK AGGREGATES LTD. v. ABBOTSFORD (MUNICIPAL DISTRICT)

British Columbia Court of Appeal; November 1989

The Municipal District of Abbotsford awarded a gravel crushing contract to a local company, even though Chinook Aggregates Ltd. was the lowest bidder. In this, Abbotsford followed its secret policy that if any local bidder was within 10% of the lowest bidder, the local company would get the contract.

This policy was deliberately not made clear to bidders because the municipality did not want to alert local contractors to the fact that they would be given preference. The invitation to bid, however, contained the typical privilege clause:

> The lowest or any tender will not necessarily be accepted.

At trial, Judge Selbie held that Contract A automatically came into existence when Chinook submitted a bid to the owner. He quoted from *Law of Contracts* by Cheshire and Fifoot:

> The normal contract is not an isolated act but an incident in the conduct of business ... It will frequently be set against a background of usage familiar to all who engage in similar negotiations and which may be supposed to govern the language of a particular agreement. In addition therefore to the terms which the parties have expressly adopted there may be others imported into the contract from its context. These implications may be derived from custom...

Basing his decision on the custom in the construction industry, Judge Selbie held that it was an implied term of Contract A between Abbotsford and the bidders that the lowest compliant bid must be accepted. The judge found that Abbotsford had breached that implied term, and he awarded damages to Chinook.

Abbotsford appealed the decision and argued that the trial judge had erred in incorporating an implied term into Contract A when that term was contrary to the express terms of the contract, namely the privilege clause. Counsel for Abbotsford referred to the 1987 decision of the Ontario Supreme Court in *Elgin Construction*, which supported his contention that custom and usage cannot override the explicit terms of a contract.

Justice Legg of the B.C. Court of Appeal refused to accept that the privilege clause gave the owner the right to exercise a local preference without revealing it in the bid documents. Where the owner attaches a condition to its offer and that condition is unknown to the contractor, it would be inequitable to let the owner hide behind a disclaimer clause. The owner was, thus, in breach of a duty to treat all bidders fairly, a duty recognized in the *Best Cleaners* case, where Justice Pratte found that implied terms of Contract A

> ... impose on the owner calling the tenders the obligation to treat all bidders fairly and not to give any of them an unfair advantage over the others.

Justice Legg added:

> ... it is inherent in the tendering process that the owner is inviting bidders to put in their lowest bid and that the bidders will respond accordingly. If the owner attaches an undisclosed term that is inconsistent with that tendering process, a term that the lowest qualified bid will be accepted should be implied in order to give effect to that process.

The court unanimously rejected Abbotsford's appeal.

---

**Qualified**: This adjective when used in connection with *bids* has two opposite meanings, depending on the context. A qualified bid may be:

- one that complies fully with the terms of the bid documents; for clarity, we shall call such bids "compliant" rather than "qualified" except when quoting from a court decision where the meaning should be clear from the context; or
- one that is in some way modified or limited so that it does not comply fully with the bid documents; this is what the word will mean in this book when applied to bids.

When applied to *bidders*, the word means something else again: A qualified bidder is, of course, one that has the qualifications and resources to do the work.

---

The Elgin and Chinook cases may not be as disparate as they first appear.

In *Chinook*, the owner discounted all local contractors by 10% before it even considered the bids. In a sense, the courts refused to let the owner act arbitrarily (although it had given itself the contractual right to do so) because it had first acted discriminatorily.

In *Elgin*, there was no secret agenda, and the valuation criteria were applied across the board. The owner was not trying to use the privilege clause to protect itself from the consequences of its own unfairness.

Thus, the difference between the decisions may be less an east-and-west division and more a court finding the application of a hidden agenda unpalatable.

Now back to Ontario, and its Court of Appeal. With barely a polite nod in the direction of British Columbia, the Ontario court chose to support the words of the bidding contract, custom or no custom.

## ACME BUILDING AND CONSTRUCTION LIMITED v. NEWCASTLE (TOWN)

Court of Appeal for Ontario; June 1992

Acme Building and Construction Limited was the lowest bidder on a project in Newcastle, Ontario. However, the owner — the Corporation of the Town of Newcastle — accepted the second lowest bid.

The minutes of the council meeting at which the decision was made showed that the successful bidder would complete the project within a shorter construction period and thus save the town some $25,000 in rent. It would also minimize disruption. Additionally, the contractor guaranteed that at least 23% of the subcontractors assigned to the project would be local. Acme guaranteed only 18% local involvement.

Acme argued in court that these considerations were irrelevant to the project as described in the bid documents.

Alternatively, it relied on expert evidence to prove that the custom and usages of the trade demanded that if such factors were to be given weight, the fact should have been made clear in the bid instructions. Finally, Acme submitted that it was the accepted custom and usage in the construction industry for the lowest qualifying bid to be awarded the contract.

The trial judge rejected the evidence that such customs and usages existed and concluded that Newcastle's procedures were fair. Acme appealed, but the Court of Appeal agreed with the findings of the trial judge.

But even if such industry customs existed, decided the court, they could not prevail over the express language of the bid documents which contained the privilege clause:

> The owner shall have the right not to accept the lowest or any other tender.

All bidders were asked to stipulate a completion date and to list their subcontractors, so Acme could not argue that these were an immaterial term of the bids. The appeal court agreed with the trial judge that the owner was not obliged to spell out the weight it was giving each of these factors.

In any case, the privilege clause gave the owner the right to reject the lowest bid and accept another qualifying bid without giving *any* reasons. The council chose to give reasons, but there was nothing in those reasons to indicate that the owner had acted improperly in making the decision, which was its contractual right to make.

The contractor relied on the decision of the B.C. Court of Appeal in *Chinook Aggregates Ltd. v. Abbotsford* for the proposition that in awarding the construction contract, an owner cannot use criteria inconsistent with the customs and usage of the trade, even if the bid documents contained the privilege clause.

The Court of Appeal was not impressed with the contractor's argument. The court held that the trial judge was right in finding that the parties' rights were governed solely by the terms of the contractual arrangements contained in the bid documents. The court, therefore, dismissed Acme's appeal.

---

*In the next case, the owner took advantage of a broadly worded privilege clause to **accept** a bid which it deemed to be "irregular" (today, "substantially compliant") but was probably "informal" (today, "non-compliant").*

## MEGATECH CONTRACTING LIMITED v. OTTAWA-CARLETON (REGIONAL MUNICIPALITY)

Supreme Court of Ontario; April 1989

In April 1986, the Regional Municipality of Ottawa-Carleton invited bids for a construction project. The Information for Tenderers required bidders to nominate each proposed subcontractor, complete an agreement to bond and submit complete documentation without any part detached.

The Information contained several important provisions:

- Failure to comply with any of the requirements in the bid documents would result in the bid being declared *irregular*.
- Bids that are incomplete, conditional or obscure, or that contain additions not called for, erasures, alterations, or irregularities of any kind, *may* be rejected as *informal*.
- The corporation also reserves the right to reject any or all bids.

Megatech Contracting Ltd. was the second lowest bidder, the lowest being Comstock International Limited. However, Comstock did not name the subcontractors it was going to use in its bid. Megatech brought this omission and other irregularities to the attention of the owner immediately after the bid had closed.

Nevertheless, the owner awarded the contract to Comstock. Megatech sued, claiming that the award to Comstock was wrongful. Megatech argued that nominating subcontractors is fundamental to the bidding process because it prevents bid shopping. Without it, the successful bidder can manipulate lower subcontract prices because having the prime contract gives it additional leverage.

Megatech claimed that the owner, when it accepted Comstock's deficient bid, put Megatech in an unfairly disadvantageous position. Had the Comstock bid been rejected as invalid, Megatech would have been awarded the contract. Megatech claimed to have suffered a loss of profit and mark-up on overhead amounting to $53,286.

Alternatively, Megatech stated that the conduct of Ottawa-Carleton entitled Megatech to claim punitive or exemplary damages in an amount equal to its loss of profit and overhead.

In its defence, the Municipality pleaded that any bidding information it distributed was primarily for the guidance of its staff and subject to amendment from time to time. The Municipality might be *guided* but not *governed* by earlier policy directives. It also strenuously contended that the awarding of the contract to Comstock was not only valid but also proper in the circumstances because of the following:

- The bid call contained clauses that gave the owner the right to reject certain bids (such as "The Council does not bind itself to accept the lowest or any tender," and "tenders which contain irregularities of any kind may be rejected as informal ...").
- Comstock's bid was substantially lower in price than Megatech's bid and complied substantially in all material respects with the guidelines.

The court had to determine whether the discretionary clauses found in the bid procedure guidelines should override the specific requirement in the bid form that subcontractors must be listed.

Said Justice Maloney:

> Really I am asked to find that the defendant [Municipality] is contractually bound to the plaintiff [Megatech] *not* to accept some other bidder's tender which is in some way deficient, that is, not strictly in accord with the bidding specifications.

The court found that the Municipality did not arbitrarily put aside the clause which said that irregular bids might be rejected. It carefully examined the bids and then relied on the general privilege clause in order to accept the bid. There may have been irregularities in the Comstock bid, but it was not invalid, and the owner was not obliged to reject it.

The question whether a failure to list subcontractors makes a bid informal (non-compliant) or just irregular (substantially compliant) is a controversial one. Owners have a legitimate interest in knowing what trades are proposed for important parts of their projects. An owner can hardly exercise its right to direct the contractor to change "subs" (or subcontractors) if the contractor fails to name them.

Based on the Information for Tenderers in the *Megatech* case, all bidders were required to name their subs.

---

> Later in this book, we will look at claims for misrepresentation in the bidding context. After all, the bid documents are the owner's "representation" of the manner in which it will run its process. Megatech might have been better off by suing for misrepresentation because, with respect to subcontractors, the owner did not conduct the process it said it would.
>
> Megatech was correct — at least in business terms — in claiming that Comstock was unfairly advantaged by its failure to list trades. While it is true that Comstock was committed to its bid price, it was not committed to its costs. Megatech was committed to its costs — principally its subtrades — when its bid was submitted. For Comstock, bid close did not mean bid closed. It was still in a position to lower its costs and thus increase its profit. Sure, Megatech may have found ways to better its costing. But, its opportunity to comparison shop ended at closing. Comstock's did not.
>
> As an aside to the main thrust of the Megatech case, we include the following quote from Justice Maloney:

> > In my view, Ron Engineering must be understood to stand for the proposition that Contract A, based on the irrevocable bid of the successful tenderer, arises only when the tender has been accepted. The mere act of submitting a tender cannot create such a contract; if it did, all responses to calls for tenders would give rise to absurd assertions of contractual rights.
>
> This appears to be a wrong reading of the Ron Engineering decision, but it is by no means unique. Not all judges find the concept of Contract A/Contract B easy to digest — see, for example, Ellis-Don Construction Ltd. v. R. in Chapter 4.

## KENCOR HOLDINGS LTD. v. SASKATCHEWAN

Queen's Bench of Saskatchewan; September 1991

In April 1984, the Government of Saskatchewan invited bids for the construction of a bridge. Kencor Holdings Ltd., a Calgary company, bid approximately $1.7 million. Graham Construction and Engineering Ltd., also a Calgary company but with a branch office in Saskatoon, bid almost $1.8 million.

The Crown's bridge branch investigated the qualifications of the bidders and concluded that Kencor was more qualified and more experienced in pier construction. It recommended that Kencor be awarded the contract.

On May 8, the recommendation and all supporting information was given to the Deputy Minister of Highways and Transportation. The following day, by way of an order-in-council, the minister awarded the contract to the higher bidder, Graham Construction.

The order stated that it was "expedient and in the public interest" to award the contract to Graham Construction, a local contractor from Saskatoon. There was no indication in the bid documents that preference might be extended to Saskatchewan bidders, and Kencor was not aware of that possibility.

However, the bid documents contained the following privilege clause:

> The Minister may refuse to accept any tender, waive defects or technicalities, or ... accept any tender that he considers to be in the best interests of the Province.

Kencor sued. It tried to prove that the Government, in the exercise of its discretion with respect to bids, was not entitled to apply a local preference policy because the bid package did not identify this as a factor which might be considered.

The court examined the body of law which had developed on the subject of bidding and its related obligations. The starting point was, of course, the *Ron Engineering* case. An implied term of Contract A imposes on the owner

> ... the qualified obligations to accept the lowest tender, and the degree of this obligation is controlled by the terms and conditions established in the call for tenders.

This decision was applied in *Chinook Aggregates Ltd. v. Abbotsford*. The court held that, by adopting a policy of secret local preference, the Municipality of Abbotsford breached its duty to treat all bidders fairly and not to give any an unfair advantage.

There was ample evidence at the *Kencor* trial that industry custom dictates acceptance of the lowest compliant bid. The court found custom to be an implied term of Contract A, which came into being when Kencor submitted its bid in response to the Government's call for bids.

"To maintain the integrity of the tendering process, it is imperative that the low qualified bidder succeed", said Justice Halvorson. This is especially true in the public sector:

> If governments meddle in the process and deviate from the industry custom of accepting the low bid, competition will wane. The inevitable consequence will be higher costs to the taxpayer. Moreover, when governments, for reasons of patronage or otherwise, apply criteria unknown to the bidders, great injustice follows. Bidders, doomed in advance by secret standards, will waste large sums preparing futile bids.

If the Government was allowed to succeed in its argument that its privilege clause was paramount, then, practically speaking, the Government would never be liable for damages in these situations.

It could always maintain that it was "in the best interests of the province" to reject the "low bid" (or lowest bid). With cabinet deliberations being secret, the public could never challenge an irrational or capricious exercise of this discretion.

"This is a blatant case of unfair and unequal treatment of [Kencor] in order to benefit another contractor", concluded the judge. He found that the Government had breached its bidding contract with Kencor and was responsible for the ensuing damages. He awarded Kencor $180,000, which represented, among other things, its loss of profit for not having been awarded the contract.

---

*The next case throws some light on a very important question: Under what circumstances can an owner reject all bids and restart the process from square one?*

## POWER AGENCIES CO. LTD. v. NEWFOUNDLAND HOSPITAL AND NURSING HOME ASSOCIATION

Supreme Court of Newfoundland; May 1991

The law of bidding and tendering, which developed almost exclusively in the arena of construction disputes, applies equally to other endeavours.

The Newfoundland Hospital and Nursing Home Association is a government funded body which operates a group purchasing programme for its member institutions. In 1988, the Association issued a bid call for certain hospital supplies. Following the closing of bids, the bid submitted by Power Agencies was found to be the "preferred bid" as defined by the provincial government's Local Preference Policy.

Soon after, the Association was contacted by another bidder which complained that Power's bid was qualified, and the bidder argued that the bid should not have been considered. The Association sought legal advice and discovered that all the bids were qualified in some way because none met the specifications contained in the bid call. The Association then cancelled the bid call and issued a new one. Power submitted a bid on the new call "without prejudice" to its right to sue under the original call. Under the new call, Power was again awarded the contract. It still sued for damages for the Association's refusal to award it the contract under the first call.

Counsel for Power argued in court that following the decision of the Supreme Court of Canada in *Ron Engineering*, it had a bidding contract (Contract A) with the Association.

In the *Ron Engineering* case, Justice Estey said that the owner had "the qualified obligation to accept the lowest tender, and the degree of this obligation is controlled by the terms and conditions established in the call for tenders". However, the Association's bid call clearly stated that "the lowest or any tender will not necessarily be accepted". The Association argued that it was not obliged to accept Power's bid or anyone else's.

Madam Justice Cameron of the Supreme Court of Newfoundland reviewed the interpretation of the *Ron Engineering* decision in the lower courts and found it inconsistent.

In Ontario, for example, in the case *Acme Building & Construction Ltd. v. Newcastle*, the court followed *Megatech Contracting Ltd. v. Ottawa-Carleton* in holding that Contract A arises only when the owner accepts the bid.

In the *Acme* case, the court also discussed *Elgin Construction Co. v. Russell Township*. In that case, the Township reserved the right to reject any and all bids, and Justice White held that no "custom of the trade" could be deemed to qualify the explicit words of the call for bid.

However, in *Chinook Aggregates Ltd. v. Abbotsford*, the Court of Appeal of British Columbia upheld a decision of the trial judge who held that in spite of the clear wording of a privilege clause, there was an implied term in the contract that the parties would comply with industry "custom and usage" regarding the rejection of the lowest bidder.

The court also held that following the decision of the Federal Court of Appeal in *Best Cleaners & Contractors Ltd.*, there was a duty to treat all bidders fairly and not to give any of them an unfair advantage. The owner in the *Chinook* case had breached this implied term of the bidding contract as well as industry custom by applying a local preference policy without letting the bidders know that such a policy existed.

Justice Mahoney in the *Best Cleaners* case said:

> The [owner's] obligation under Contract A was not to award a contract except in accordance with the tender call. The stipulation that the lowest or any tender need not be accepted does not alter that.

Justice Cameron decided that the cases mentioned above did not go as far as to say that the owner, having clearly stated that the lowest or any bid need not be accepted, was obliged to award a contract if there was a bid complying with the terms and conditions of the call.

The courts used the implied term that all bidders be treated fairly to ensure that what is awarded is what was called. In other words, if a contract is to be awarded on anything *other* than the commonly accepted basis, this must be revealed to all bidders.

"In this case", said the judge, "all bidders were aware of the terms and conditions set out in the call for tenders and that the provincial preference legislation would be applied to assess the tenders submitted". Therefore, the Association was not obliged to award a contract, even if Power met all the requirements set out in the bid call.

Power also argued that the Association had acted unfairly by failing to award it the contract and subsequently issuing a new and only slightly different call for bids after all competitors knew of Power's bid on the first call. The court rejected this argument for two reasons.

First, there was no evidence of bid-rigging or bad faith on the part of the Association. After obtaining advice that none of the bids met the specifications, the Association chose to re-tender. "In the circumstances, a reasonable course of action", decided the court.

Second, the requirement of fairness to contractors under public law as well as any such term implied by contract was meant to ensure procedural fairness, but not necessarily a fair outcome. Power's claim, therefore, was dismissed.

---

*Such emphasis on good faith and fairness in what might otherwise be a somewhat troubling decision is reassuring. Power was justified in complaining when the owner called a re-tender on the very same project or product after all bidders had seen the prices bid by the competition.*

*If owners, especially public owners, are permitted to use the privilege clause to establish pricing for their projects and then to either negotiate or re-tender into an auction, the integrity of the bidding system, a delicate thing at the best of times, will be severely jeopardized.*

*The next case, although not a construction case, represents a strong affirmation of the privilege clause.*

## MARTSELOS SERVICES LIMITED v. ARCTIC COLLEGE

Court of Appeal of the Northwest Territories; January 1994

This appeal has little to do with construction, but the issues raised are important. They touch on the duty of fairness in the bidding process and the effect of the so-called privilege clause.

In May 1990, Arctic College advertised for a contract for janitorial services at its facility in Fort Smith, N.W.T. The call for bids contained the privilege clause:

> The lowest or any tender not necessarily accepted.

Only two companies submitted bids: Martselos Services Limited and P&A Security and Safety.

After it opened the bids but before it awarded the contract, Martselos complained to Arctic College that one of the college's employees was a director and shareholder of P&A. The college, nevertheless, awarded the contract to P&A, which had submitted the lowest bid.

Martselos sued the college, claiming damages for breach of Contract A. It complained that the college had not acted fairly in awarding the contract: the employee had a conflict of interest and also could have exploited inside information in preparing the bid. The college disputed this claim.

The trial judge found in favour of Martselos and awarded it damages in the amount of its anticipated profit from the contract for janitorial services. He held that the college had a duty to act fairly to all bidders, and that included an obligation to avoid even the suspicion of any unfair advantage being enjoyed by its employees.

He also held that the employee's failure to disclose the potential conflict of interest was a sign of bad faith, which disqualified the competitor from consideration as a bidder. This left Martselos as the only eligible bidder, and the court decided that the college should have awarded the contract to Martselos.

Arctic College appealed. The Court of Appeal of the Northwest Territories decided that the judge had erred in his judgment. The critical factor in its decision was the privilege clause.

In *Ron Engineering*, Justice Estey of the Supreme Court of Canada said:

> … the integrity of the bidding system must be protected where under the law of contracts it is possible so to do.

"In my opinion", stated Justice Vertes speaking for the Court of Appeal, "this should be considered as a duty to treat all bidders equally but still with due regard for the contractual terms incorporated into the tender call".

But even if P&A did breach the conflict of interest guidelines, that did not mean that the college acted in bad faith in awarding the contract. Only the bidder showed bad faith.

The effect of this bad faith could only be the disqualification of P&A's bid. It could not create an obligation on the part of the college to Martselos. The college was not obliged to award the contract to any bidder, as the privilege clause made quite clear.

When Martselos submitted its bid, its action gave rise to Contract A. One of the explicit terms of the contract in this case, said Justice Vertes, was that the lowest or any tender would not necessarily be accepted.

Martselos argued, however, that it was the industry practice or custom to award such contracts to the "low bidder" (or lowest bidder). Canadian jurisprudence, retorted the judge, does not recognize any precedence of industry custom over explicit terms of a contract, except in special circumstances.

He quoted *Elgin Construction* in which the court dismissed a claim based on this argument and held that the explicit words of the privilege clause contained in the Instructions to Bidders governed. He added that the same reasoning was applied in a number of other cases, with a similar result — the plaintiff was not successful. (All the cases he mentioned are summarized above.)

Practice and industry custom have been held to override the privilege clause only in special circumstances, continued the judge. Those circumstances have been found where:

- an owner, in awarding the contract, had relied on undisclosed criteria;
- the owner took into account irrelevant or extraneous considerations;
- the bid call contained specific provisions that were inconsistent with the general privilege clause; or
- the whole bidding process was a sham.

Examples of such special situations were *Chinook Aggregates* and *Best Cleaners*.

In the *Martselos* case, the court saw no special circumstances that would warrant interfering with the explicit terms of the bid call. Justice Vertes added:

> Even if it can be said that the industry practice was to award the contract to the lowest bidder — and I am far from convinced that the evidence reveals such "practice" since there were occasions when the work was not done by the low bidder — there was no evidence that the [college] had done anything in assessing the two bids so as to jeopardize the integrity of the bidding process.

There was nothing in the bid documents to limit the scope of the privilege clause. The college was not applying undisclosed criteria nor was it giving some advantage or preference to the low bidder because of the employee's position. There was no evidence that the college used different standards in assessing the two bids or that the bidding process was a sham.

Furthermore, even if the court were to assume that the college should have eliminated the competing bid, leaving only Martselos as an eligible bidder, the owner could still rely on the privilege clause to reject it.

The fact that Martselos was the only eligible bidder made no difference — the college could have decided not to award the contract at all. The privilege clause, concluded the judge, is a complete answer to the bidder's claim.

---

*What should an owner do when it receives a single bid on a project? The City of Windsor, Ontario, took the wrong steps and paid for its mistake.*

# COLAUTTI BROTHERS MARBLE TILE AND CARPET (1985) INC. v. WINDSOR (CITY)

Ontario Court (General Division); December 1996

In 1990, the City undertook renovations and additions to a convention centre which it owned and operated. Colautti Brothers Marble Tile and Carpet (1985) Inc. submitted a bid for the flooring package. It was the only bid submitted, but the employees of Colautti that were present at the opening of bids were not aware of that. The City opened the bid and found that Colautti's bid was significantly over the City's budget. The bid price was disclosed in public.

The City then entered negotiations with Colautti but failed to get an acceptable price. Its budget for the flooring was $700,000; following negotiations, Colautti's initial bid of $977,322 was lowered to $834,250.

The City did not communicate its dissatisfaction to Colautti. In fact, it never indicated in any way that Colautti's bid was either accepted or rejected; it simply did not enter into a formal contract with Colautti.

Instead, it decided to re-tender the flooring package by breaking it up into three smaller packages. This time, the result was more satisfactory. The City received low bids in the total amount of $707,244 for the work (on which Colautti had previously bid). Colautti did not bid on any of the new flooring work. Instead, it took the City to court. It asked for damages for breach of Contract A. Contract A had arisen when Colautti submitted its bid in response to the first bid call.

There were several issues before the court:

- Was the City entitled to rely on the privilege clause in which it reserved the right to accept or reject any and all bids?
- Assuming that the City owed a general duty to treat all bidders fairly, was the City obligated to accept the lowest bid it received, even though it was the only bid received?
- If the answer to the second question was "No", was the City, nevertheless, in breach of an implied duty of fairness?
- Is the City liable, and, if so, what is the measure of Colautti's damages?

The court found that following the decision in *Ron Engineering*, a line of cases has developed which suggests that once Contract A comes into being,

- the law imposes rights and obligations on the parties that are consistent with the protection and promotion of the integrity of the bid system where, under the law of contracts, it is possible to do so;

- the owner owes a general duty to treat all bidders fairly;
- the owner has the right to include in the bid documents stipulations and restrictions and to reserve privileges to itself; and
- general custom in bidding, and particular local customs, can result in implied contractual rights.

An expert testified for Colautti that based on his experience, the City had always adhered to a policy of awarding contracts to the lowest qualified bidder. He expressed the opinion that if the City had not intended to award the contract to a sole bidder, it should have returned Colautti's bid unopened and then re-tendered the work. When the City opened Colautti's sole bid, Colautti was placed at a significant competitive disadvantage because its unit prices for certain items of work were disclosed to the public.

The City's policy was, indeed, to award contracts to the lowest qualified bidder. Colautti had been prequalified. In the circumstances, argued Colautti, the general custom of the City and its policy to award contracts to the lowest qualified bidder should override the privilege clause that otherwise entitled the City to accept or reject any and all bids.

Counsel for the City argued that the formation of a contract between Colautti and the City would have arisen only if the City had actually *accepted* Colautti's bid.

In addition to the standard privilege clause, the bid documents stated (emphasis added):

> It shall be understood by all bidders that the Tender shall be valid and *subject to acceptance* by the City.

Colautti was a contractor with extensive experience in bidding on jobs. The bid documents made it patently clear that the City reserved the right to reject any or all bids, that it was not under any obligation to accept a sole bid and that Colautti could only expect to receive a contract if the City accepted its bid. Since it was clear that the City did not accept Colautti's bid, the argument concluded, the action should be dismissed.

Furthermore, counsel for the City submitted that the conduct of the parties after Colautti's bid was opened was significant. Colautti's bid was much higher than the budget the City had established for the work. In the days following the opening of the bid, the City's project manager twice invited Colautti to submit prices in areas of the proposed work where the City felt that cost savings might be achieved. Colautti responded with price quotations on both occasions. It did not object in any way to the supposed lack of fairness with which it was being treated.

An analysis of the issues, said Justice Valin of the Ontario Court, must begin with an understanding of the specific legal nature of a bid. A bid is simply an offer to carry out the work specified, subject to the terms and conditions stated and at the price quoted. As such, it is subject to all the rules which apply to offers in general.

He quoted from *Goldsmith on Canadian Building Contracts*:

> If the invitation to bid contains specific terms and conditions, *e.g.* not to revoke the tender for a specified period, such conditions will form part of contract A, and can be enforced in the same way as any other contractual obligations.
>
> Insofar as contract B is concerned, however, the tender is still only an offer which may or may not be accepted by the owner. Contract A is not an agreement to enter into contract B, and unless and until the tender is accepted by the owner, no contract B will ever come into existence.
>
> Whether, in any given situation, a contract A ever comes into existence necessarily depends on the particular facts of each individual case. If it does, it is difficult to see on what basis an express term of the contract, *e.g.* a provision in the invitation to bid that the owner is not obliged to accept the lowest or any tender, can be displaced by an implied term or by a custom of the trade to the contrary.

In Ontario, the law is clear that any custom of the trade is subject to the explicit words of the bid documents. In *Acme Building*, the Ontario Court of Appeal held that industry custom could not prevail over the express language of the bid documents.

"I am of the view", said Justice Valin, "that the decision of the Ontario Court of Appeal in Acme Building is the law in Ontario. I am bound to follow that decision".

He, therefore, found that the City was entitled to rely on the privilege clause and that industry custom could not prevail over the express language of the privilege clause, which was clear and unequivocal. Once Colautti's bid was opened, the City acted properly in trying to achieve cost savings. While the City owed a duty to bidders to act fairly, it also had the duty to its taxpayers to act in a financially responsible manner.

The City was entitled to decide that it was not going to accept Colautti's bid. The fact that the City did not formally reject Colautti's bid was of no consequence. The City was not obligated by an implied duty of fairness to accept Colautti's bid, even though it was the only bid received.

But did the City treat Colautti fairly when it opened its bid, knowing that it was the only one received?

The CCDC (Canadian Construction Documents Committee) *Guide to Calling Bids and Awarding Contracts*, which is widely used in the construction industry, has the following comments regarding single bids:

If only one bid is received and the owner, under these circumstances, is unable or unwilling to award a contract, the bid should be returned unopened to the bidder. If the owner is not so constrained, a meeting with the bidder should be called by the consultant.

Any rules governing the opening of the bid and subsequent negotiations prior to contract award should be agreed on at such a meeting. At this stage, failing agreement, the bidder may choose to withdraw its bid unopened, and it should be allowed to do so.

On the closing date, the City was aware that Colautti's bid was the only bid received. It knew that in addition to requesting a stipulated price for doing the work, the bid documents required bidders to disclose a number of unit prices associated with doing the work.

The court found that the City should have realized that by opening Colautti's bid without prior agreement, the City placed Colautti in a potentially serious disadvantage with its competitors in the event it was not awarded a contract. The City ought to have followed the procedure outlined in the CCDC Guide.

The court found that the City had breached its duty of fairness to the bidder by

- failing to advise Colautti that only one bid had been received;
- failing to try to negotiate with Colautti the rules under which its bid might be opened; and
- failing to offer to return Colautti's bid unopened.

Accordingly, Colautti was entitled to recover from the City damages that had arisen from that breach.

When it came to assessing damages, the court found that several factors worked in the City's favour. Following the opening of Colautti's bid, the City made an attempt to negotiate a price with the bidder in good faith. Colautti did not submit a bid on any of the three packages re-tendered by the City because it was busy with other work at the time. The City did not use Colautti's bid to "shop prices".

For these reasons, Justice Valin restricted Colautti's damages to the reasonable cost of preparing and submitting its original bid, together with the additional costs related to quoting on the cost-saving measures.

---

*In some respects, fairness is in the eye of the beholder, and here, the beholder is a judge sitting in a court, hearing conflicting evidence. Had the contractor been given the opportunity to take back its bid unopened because it was the only bid, common sense says the contractor would have rolled the dice and* negotiated with the City. Perhaps the City withheld the information to keep some tension in the negotiations. In this case, whether it was intentional or not, the court found that the failure to disclose was unfair.

## THOMPSON BROS. (CONST.) LTD. v. WETASKIWIN (CITY)

Alberta Court of Queen's Bench; August 1997

The City of Wetaskiwin called bids for the excavation of clay from a lake. The bid documents included several privilege clauses. First, the standard one:

> The lowest or any Tender may not necessarily be accepted.

Then the same with frills:

> The Owner reserves the right to reject any and all Tenders or to accept the Tender deemed most favourable in the interests of the Owner.

And, finally, presumably in case the bidders failed to notice the first two:

> The Owner reserves the right to reject any and all Tenders; the lowest will not necessarily be accepted. Without limiting the generality of the foregoing, any Tender may be rejected which:
> 
> (a) is incomplete, obscure or irregular;
> (b) has qualifications unacceptable to the Owner;
> (c) omits a bid on any one or more items in any Schedule to the Form of Tender; or
> (d) is accompanied by insufficient or irregular bond or cheque.

The bid documents also included a clause which reserved the City the right to let other contracts to other contractors.

When bids closed, Thompson Bros. (Const.) Ltd. was the lowest. After a review of the bid, the City Manager recommended to the City that Thompson be awarded the contract. At the council meeting, where the decision was to be made, that recommendation was tabled for the reasons that follow.

While bids for the lake excavation were outstanding, the City was planning a landfill site which required a clay lining and berms. The City thought that the excavated material from the lake might be used at the landfill site. The City asked Thompson and the second low bidder on the lake excavation (Central Oilfield Services) to provide a price for the landfill work.

Thompson believed that the pricing for the landfill work was separate from the excavation work already tendered. Central knew that the City's intention was to award a contract for lake excavation and for the landfill site to the lowest *cumulative* bidder.

On the original bid, Thompson was some $31,000 lower than Central. After pricing for the landfill work was received, the aggregate of Thompson's two prices was $2,000 higher than the aggregate of the two prices received from Central. The City combined the lake and landfill work and awarded a single contract to Central, which left Thompson in the mood to sue.

The City believed that if it was protected by the privilege clause, it was safe to make the award as it saw fit. Thompson argued that it was a custom of the trade for the lowest bidder to be awarded the work. After considering a variety of authorities, including *Ron Engineering*, *Elgin*, *Acme* and *Chinook*, the court ruled that the City was in breach of Contract A. The following passage from the reasons is instructive:

> The problem in this case is the use to which the City of Wetaskiwin put the tenders it received and in particular [Thompson's] tender and that of Central. Rather than accepting or rejecting one tender or another, or accepting none of those submitted, which it had the right to do, the City had [Thompson] and Central bid on entirely different work, the Landfill project.
>
> When they received those bids, the City coupled them with the tender submitted with respect to the Lake project and awarded the contract to Central who had the lowest cumulative price. The contract awarded was not responsive to the tender process. Rather, it was for work of a different scope than that contemplated in the tender documentation.
>
> What the City did also amounted to a change or modification of the scope of work after the close of tenders. The City as well changed the use to which the tenders were to be put by using them as a component with the Landfill project price to determine who would be awarded the contract. By using the [Thompson's] tender in this manner, the City gave Central a second chance to bid on the Lake project which was akin to a form of bid shopping and was unfair to [Thompson].
>
> It was not a term of the tender documentation that the tenders could be used in this manner and by doing so the City was in breach of Contract A. Where there has been such a breach or breaches, it would be unfair to allow the owner to use a privilege clause as a shield when the other contracting party seeks a remedy, and to override the implied term that the lowest qualifying bid will be accepted where there is no reasonable cause for rejecting it. The evidence disclosed that there was no cause for rejecting [Thompson's] tender ....

In spite of the City's efforts to make this a privilege clause case, the court refused to go along. To do so would allow the City (and presumably any other bid calling authorities) to use the privilege clause to avoid *ever* being in breach of Contract A.

The decision in *Thompson* was driven by the factual reality that the work that was awarded was not the work that was bid on. What raised the court's antenna was the unfairness of the City's action, which amounted to a form of bid shopping. Central knew Thompson's price for the lake excavation. It also knew that the lake excavation and landfill work would be lumped together and awarded to the best cumulative bid.

Curiously enough, had the City awarded the lake excavation to Thompson and the landfill to Central, it would have saved over $30,000 on what it ultimately paid to Central — never mind what it paid in the trial award to Thompson.

Once the court found the City liable for breach of Contract A, it determined that Thompson's recovery should not be limited to the cost of its bid preparation. The court awarded Thompson its estimated profit on the lake excavation work ($3,200 for bid preparation versus $88,000 for lost profit).

The judge approached the damages issue in two steps. First, he had to decide, all things being equal, whether the lake excavation work would have been awarded to Thompson. On this score, the judge concluded that it was a "virtual certainty" that Thompson would have been awarded the work, based on the recommendations of City staff. This outcome was deflected when the City decided to entertain doing the landfill project at the same time.

The judge then looked at Thompson's claim for $88,000 as the contribution it would have enjoyed from executing the work. The judge reviewed several other cases on the point and approached the assessment of damages on the basis of lost opportunity. He looked at cases where other judges had applied discounts or contingency factors to reflect the uncertainty of construction work. He also looked at whether an award of some or all of the lost contribution would be too remote. After considering the facts and the reasoning of fellow judges, the judge determined that the award to Thomson was a near certainty requiring no discount. The full $88,000 was awarded.

---

*The reasoning in Thompson foreshadowed the outcome in the M.J.B. case at the Supreme Court of Canada where, upon finding a high probability that the wronged contractor would have received the work but for the owner's wrongful behaviour, the full contract profit was awarded.*

*In the last year of the millennium, after many years of scattered decisions, the Supreme Court of Canada finally had a go at the privilege clause.*

## M.J.B. ENTERPRISES LTD. v. DEFENCE CONSTRUCTION (1951) LTD.

Supreme Court of Canada; April 1999

When M.J.B. was granted leave to appeal to the Supreme Court of Canada, it gave rise to considerable speculation. Would the Supreme Court ratify the lower courts' notion that a duty of fairness should be implied in bid documents? Would the reign of *Ron Engineering*, the Queen Mother of bid cases, survive a review by the Supreme Court? Would the rights conferred on an owner by the so-called privilege clause be settled once and for all? The *M.J.B.* decision supplied the answers — sort of.

Defence Construction invited bids for the construction of a pump house. The work included excavation and placing of water pipe and backfill in trenches. The original bid documents provided that at Defence's option, the backfill over the pipe could be gravel, native backfill or slurry concrete. Separate unit prices for the three types of backfill were required.

By addendum, Defence required each bidder to submit a single unit price which would apply for all three types of backfill. It would be interesting to see what the bidders would do with this bit of pricing strategy by Defence.

Four bids were submitted. The low bid was from Sorochan Enterprises Ltd., which had added the following note to its bid:

Please Note:

Unit prices per metre are based on native backfill (Type 3). If Type 2 material is required from top of pipe zone to bottom of surface material for gravel or paved areas, add $60.00 per metre.

The second low bidder was M.J.B. It and other bidders complained that the Sorochan note was a qualification. Defence judged Sorochan's note to be only a clarification and awarded Sorochan the contract, which Sorochan carried out. M.J.B. sued Defence for breach of Contract A.

Not surprisingly, the Contract A and Contract B paradigm from *Ron Engineering* made feature appearances at trial, in the Alberta Court of Appeal and in the Supreme Court of Canada. The rights and obligations attached to the privilege clause were the focus.

At trial in the Alberta Court of Queen's Bench, the judge held that Sorochan's bid was invalid because the note constituted a qualification. However, the judge went on to find that Contract A between Defence and M.J.B. never came into existence.

From there, the judge went on to make a paradoxical decision. On the one hand, M.J.B. had its action dismissed with costs. On the other, the judge directed Defence to pay all bidders the expenses of bid preparation:

Because the tender submitted by Sorochan and accepted by the [Owner] was a qualified tender and not a valid tender, it seems to me that the [Owner] may have been in a technical breach of [its] obligation to treat all of the tenderers fairly, and consequently the other tenderers should be reimbursed for the expenses incurred in the preparation and submission of their respective tenders.

M.J.B. sought leave to re-argue parts of its case in front of the trial judge. It lost the application. However, during the leave hearing, the judge said he had been mistaken in finding that Contract A had not come into existence.

M.J.B. appealed to the Alberta Court of Appeal. There was no doubt in the court that Contract A came into being with M.J.B. The appellate court turned its attention to a term of Contract A — the privilege clause:

> The lowest or any tender shall not necessarily be accepted.

The appellate court dismissed M.J.B.'s appeal and included in its reasons the following brief statement:

> It is the view of this panel that the privilege clause... is a complete answer to M.J.B.'s action.

Curiously, the Alberta Court of Appeal upheld the trial judge's decision that the other bidders were not treated fairly. However, in affirming this finding, it limited recovery to M.J.B. The following is the language of that affirmation:

> We are, in the circumstances, however, prepared to affirm the Trial Judge's recommendation that fairness dictates that M.J.B. be reimbursed for the provable costs of preparing its rejected tender, although these costs were not specifically pleaded in this action.

M.J.B. was successful in getting leave to appeal to the Supreme Court of Canada. The following three questions were put to the Supreme Court of Canada:

> To determine if ... the privilege clause:
>
> (i) allows a person calling for bids to completely disregard the lowest proper and acceptable valid bid and to award the contract to anyone, including a non-compliant bidder.
> (ii) allows the person calling for bids, an absolute and unfettered discretion in awarding the contract.
> (iii) allows the person calling for bids to then commence bid shopping with contractors submitting bids and contractors not submitting bids.

M.J.B. was successful at the Supreme Court of Canada and was awarded the profit it expected to earn in carrying out the contract.

The M.J.B. bid gave rise to a Contract A, which imposed obligations on Defence.

When the bid documents expressly state or imply as a term that only valid bids will be accepted, the privilege clause cannot override the owner's obligation to accept only compliant bids. The privilege clause is only one of a number of clauses in the bidding contract and must be read "in harmony with the rest of the bid documents".

The award to Sorochan, a non-compliant bidder, was a breach of Contract A. On the balance of probabilities, had Defence not breached Contract A, the work would have been awarded to M.J.B.

The trial court and the Alberta Court of Appeal in *M.J.B.* held that the privilege clause was a complete answer to M.J.B.'s complaint. This finding was the complete opposite of the decision in *George Wimpey* (now renamed *Tarmac*) *Canada Ltd. v. Hamilton-Wentworth (Regional Municipality)*. In *Wimpey*, the owner thought the privilege clause was a complete answer and rejected Wimpey's low bid without giving reasons. That court held that there needed to be a good business reason for rejecting the low bid.

In *M.J.B.*, Defence invoked the privilege clause to award the contract to a non-compliant bidder. The Supreme Court of Canada held that that was wrong, based on its analysis of all of the bid documents:

> However, the privilege clause is only one term of Contract and must be read in harmony with the rest of the Tender Documents. To do otherwise would undermine the rest of the agreement between the parties....
>
> Therefore I conclude that the privilege clause is compatible with the obligation to accept only a compliant bid. As should be clear from this discussion, however, the privilege clause is incompatible with an obligation to accept only the lowest compliant bid. With respect to this later proposition, the privilege clause must prevail.

With those two statements, contractors, in general, had won and lost. M.J.B. had argued that the bid documents obliged Defence to award the contract to the lowest compliant bidder. The Supreme Court of Canada found that the obligation on Defence was to accept a compliant bid but not necessarily the lowest one. In the M.J.B. bid documents, the privilege clause had some teeth but not a full set.

The Supreme Court ruled that an owner needed a good reason to invoke the privilege clause to bypass the low bid — like a "nuanced" view of cost.

The court awarded M.J.B. damages for loss of profit on the construction contract it should have won.

---

*In the case law leading up to M.J.B., several trial judges attempted to send a message to owners (and to contractors in subtrade situations) that they needed to clean up their bid behaviour. These judges used contract theory, tort theory and even unjust enrichment. They implied a duty of fairness as a counter balance to the privilege clause to prevent it from being a carte blanche for a bid authority to do whatever it pleased. With all these decisions in the background, the industry looked to the M.J.B. case for unifying direction.*

*Perhaps, curiously, the M.J.B. decision concentrated on the rules of contract construction rather than embedding some principle of fairness to be applied in bidding situations. That the notions of fairness being advanced by the lower courts were neither approved nor rejected may be the most surprising non-component of the judgment.*

*Neither the trial nor the appellate decision mention the term "bid shopping", yet a question about that practice was included in the questions submitted to the Supreme Court of Canada.*

*When a court has a question put to it that is not germane to its decision in the case at hand, any pronouncements on that question are not binding on lower courts, although they may have some persuasive value. In the Supreme Court of Canada decision in M.J.B., bid shopping is not a necessary part of the ultimate decision. The term is used once, and then used in an excerpt from the trial evidence of a representative of Defence. The quote appears to have been made to illustrate the point that the bid documents issued by Defence did not contemplate post-bid negotiations between Defence and all the contractors that submitted bids.*

*In the final analysis, the Supreme Court of Canada did not deal with bid shopping because the M.J.B. case did not require it. Until recently, the industry view of bid shopping depended on whether the industry segment was the shopper or the shoppee. Supreme Court of Canada appears to have resolved this ambivalence in the Naylor case which follows later in this book. Among other things, the Naylor decision means that bid shopping is a breach of Contract A between the contractor and the trade it has "carried" in its prime bid.*

*Some thought that the Supreme Court of Canada would use M.J.B. as an opportunity to overturn Ron Engineering. When one looks at the questions put to the court on the appeal, it is apparent that a full frontal assault on that decision was never in the cards.*

*Ron Engineering was preoccupied with an invisible error in a contractor bid. The pre-Ron Engineering case law held that once an owner knew there was a mistake, the owner could not then accept the offer. The problem that existed then was how to avoid frustrating the whole bid process by permitting a contractor to yell mistake and turn its irrevocable bid (and accompanying bid security) into a no-recourse, revocable bid.*

*Ron Engineering resolved this tension (sort of) by setting up the Contract A/Contract B paradigm and deciding against the mistaken contractor. The industry's grasp of this proposition is signalled by the relatively small number of cases coming to court on invisible mistake by the contractor.*

*The Contract A/Contract B paradigm disposed of the invisible mistake issue but at the same time created a vacuum. Every contract has consideration and inter-party obligations. In the discussion of the mutual obligations under Con-*

tract A, Justice Estey referred to the owner's "qualified obligations" to award Contract B to the low bidder. He then stated that the extent of such "qualified obligations" was "controlled" by the bid documents.

As a result, the vast majority of the case law invoking Ron Engineering as authority is not on invisible mistake at all. The case law is concerned with the terms of Contract A, be they express or implied, and whether or not one party or the other has breached them. Privilege clauses are part of that enquiry. In M.J.B., the court focused, first, on whether Contract A came into being and, second, on how that particular privilege clause fit into that particular set of bid documents.

The principal findings in M.J.B. were that Contract A did come into existence (with all bidders except Sorochan) and that the owner's "qualified obligation" included that it would not award the work to a non-compliant bidder. The notion that Contract A included an implied positive obligation to award to the low bidder was specifically rejected.

## TARMAC CANADA INC. v. HAMILTON-WENTWORTH (REGIONAL MUNICIPALITY)

Court of Appeal for Ontario; September 1999

The main issue in this case (the trial decision of which case was cited as *George Wimpey Canada Limited v. Hamilton-Wentworth (Regional Municipality)*) was the meaning and scope of the so-called privilege clause in bid documents.

The Regional Municipality of Hamilton-Wentworth issued a call for bids to construct a road. Only prequalified contractors could submit bids. The bids were based on unit prices for the areas and quantities stipulated in the bid documents.

Thus, the only differences between the various bids were in the unit prices and the totals resulting from multiplying the given areas and quantities by those prices. There was no scope for variation of the work, the method of doing it or the nature or quantity of the materials to be used.

The privilege clause in the bid documents provided:

> The Region reserves the right to reject any or all bids submitted or any part and the lowest or any bid will not necessarily be accepted.

There was no indication in the bid documents of any policy or criteria to be used in exercising the privilege.

George Wimpey Canada Limited (now Tarmac Canada Inc.) submitted a total price of $8,507,000, the lowest of the six bids. The second lowest bid was that of Dufferin Construction Company, which was approximately $20,000 higher.

The Region's Director of Roads, the official responsible for ad[ministering the] bidding process and analyzing the bids, recommended that t[he contract be] awarded to Tarmac as the lowest bidder.

However, Dufferin was a local contractor; it paid large sums in realty and business taxes to the Region. It employed over 250 workers that lived in the area. At the time, most of them were laid off due to lack of work. Dufferin had worked on a number of projects for the Region before; Tarmac had not.

Dufferin lost no time in pointing all this out to the Region, adding that there would be no repercussions if the contract were awarded to other than the lowest bidder. The Region gave in and awarded the contract to Dufferin without giving a reason.

Tarmac sued. It argued that the owner has a duty to treat all bidders fairly and in good faith. Therefore, the owner must accept the lowest bid subject only to qualifications stated or implied in the bid documents regardless of the privilege clause.

Tarmac relied on *Chinook Aggregates v. Abbotsford* to show that the owner does not have the right to rely on undisclosed terms and conditions. The principle of fairness was clearly stated in the decision of the Supreme Court of Canada in *Best Cleaners* and applied in *Kencor Holdings Ltd. v. R.*, where the judge stated:

> To maintain the integrity of the tendering process it is imperative that the low, qualified bidder succeed. This is especially true in the public sector. If governments meddle in the process and deviate from the industry custom of accepting the low bid, competition will wane. The inevitable consequence will be higher costs to the taxpayer. Moreover, when governments, for reasons of patronage or otherwise, apply criteria unknown to the bidders, great injustice follows.

The principle was also followed in several other cases.

The Region, on the other hand, took the position that the privilege clause was plain and unambiguous. It allowed the owner to reject the low bid without being obliged to give its reasons for doing so. This was made clear in the decision of the Ontario Court of Appeal in *Acme Building & Construction Ltd. v. Newcastle (Town)*:

> ... even if there was acceptable evidence of custom and usage known to all the tendering parties, it could not prevail over the express language of the tender documents which constituted an irrevocable bid once submitted, and a contract when and if accepted.

The court in the *Acme* case decided that the privilege clause

> ... gave the [owner] the right to reject the lowest bid and accept another qualifying bid without giving any reasons.

In that case, there was a privilege clause very similar to the one in *Tarmac*'s case, but there was no undisclosed policy of preferring local contractors. The owner accepted a bid that provided for an earlier completion date than the low bid and included a greater number of local subcontractors. The bid documents requested information on completion periods and local subs.

The Region argued that the privilege clause was broad enough to allow rejection of the lowest bid for any reason other than fraud, including favouritism to a local contractor. The regional council should be answerable only to its voters when it rejects the lowest bid, and the court should not interfere with such a political decision.

Justice Cameron of the Ontario Court found the words of the privilege clause in the contract to be short, simple and understandable. He found no custom of the trade which could modify them. The owner, therefore, could accept or reject any bid and need not give reasons for rejecting the lowest bid.

However, Justice Cameron found that the law implies an obligation of fairness when the owner exercises its rights under the privilege clause. When the Supreme Court in the watershed *Ron Engineering* decision laid down the law of bidding and tendering, it emphasized the need to protect the integrity of the bidding system. This means that the owner must ensure that all bidders bid on the same basis without hidden preferences.

The language of the privilege clause was clear, but not clear enough: more explicit language would be required to exclude the implied obligation of fairness and good faith. Moreover, if there is any doubt, the rule of *contra proferentem* can be applied to interpret the clause against the party that drafted it. The court concluded:

> In the circumstances of this case, the [general contractor] has established on the balance of probability that the [owner] did not act fairly or in good faith when it awarded the contract to a tenderer who was not the low bidder, and no basis of possible decision on other grounds was disclosed in, nor can one be implied from, the tender documents or any published policy of the [owner].

This finding does not strip the privilege clause of all meaning, added the judge. It still permits the owner to reject the low bid in the case of some *force majeure*, or if it decides not to proceed with the project because the bids are above budget or if changed circumstances negate the viability of the project or adversely affect the low bidder's qualifications. It would also permit rejection based on a pre-published policy.

In coming to his decision, the judge took into account the fact that the owner did not give any reasons for its decision. He accepted that Acme gave the owner the right not to give reasons, but added:

> ... in the absence of possible grounds in the tender documents or a published policy ... [the owner] risks an adverse finding of fact by the court.

The court awarded Tarmac damages amounting to the difference between the revenue it would have received had it been awarded the contract and the costs it would have incurred in performing the work if economically and skilfully done. The Region appealed.

In a unanimous judgment, the Court of Appeal for Ontario upheld the decision of the trial judge. In *Acme*, the owner relied on the privilege clause in awarding the contract to a bidder that was not the lowest, but it gave reasons for doing so. These included the earlier date for completion specified by the successful bidder and the costs that would save the owner. The completion date was a material term specified by the bid documents, and the owner was entitled to base its decision on that element of the bids. There was no breach of the duty to act fairly.

The court rejected the Region's argument that the trial judge was wrong in taking into account the Region's failure to give reasons. The owner did not have to give reasons, but in not giving them took a risk. In view of the circumstances surrounding the award process, decided the court, the judge was justified in finding a breach of the owner's contractual obligation to treat Tarmac's bid fairly.

The court dismissed the Region's appeal, with costs.

## SOUND CONTRACTING LTD. v. CITY OF NANAIMO

Court of Appeal for British Columbia; May 2000

In June 1990, Sound Contracting Ltd. submitted a bid to the city of Nanaimo, British Columbia, for a project known as Hammond Bay, and was the low bidder.

Prior to this bid, Sound had carried out the Terminal Avenue Project for the City. At bid, Sound was low but the project was over Nanaimo's budget. Nanaimo cut work to get within budget and Sound claimed for its overhead and related charges for the deleted work. When Nanaimo balked, the claim went to arbitration. The arbitration decision was under reserve when the Hammond bid was submitted on June 12.

On June 15, the arbitrator delivered its decision on Terminal Avenue and awarded Sound approximately $8,000 in relation to the deleted work. The total award cost Nanaimo an additional $22,600 on a final contract price of $180,000.

With Hammond Bay over budget and the contract containing the same arbitration provision, the City concluded it would not get a satisfactory (to it) credit from Sound if work was deleted after the contract was awarded. When analyzing Sound's bid, the City therefore assumed it would have to hire an engineer to monitor Sound and factored that cost into the analysis of the bid. It also factored in an amount it might have to pay for anticipated work deletions, and the likely costs of legal, staff and arbitration expenses.

On the analysis of the second lowest bid, neither the engineer nor the cost of deletions was factored in. The staff determined that the contractor's work could be supervised by the City's own employees, and that that contractor had never claimed for contract deletions or made any previous claims resulting in extra arbitration costs.

Sound's bid of $540,639 was the numerical lowest bid for the Hammond Bay project. The second low bid was $573,330. Nanaimo's staff concluded that awarding the contract to Sound "may in fact not be the lowest overall cost to the City." In his report to the City, the staff reported:

> It is staff's conclusion that to award the tender to the low bidder, Sound Contracting, would result in a cost to the City in excess of the second low bid ... Most significantly, if the contract were awarded to Sound it would be imperative based on the City's past experiences with this particular contractor, that an independent, qualified inspector be hired to supervise the work on a daily basis. This would add to the project costs...

City Council concluded that the second low bid was the most favourable to the City, and the one which would result in the best overall value. Accordingly, the contract was awarded to the second low bidder. Sound sued.

The trial judge held that it was unfair of Nanaimo to evaluate the two bids differently. She found that Nanaimo imported undisclosed criteria in ruling against Sound's bid and was, therefore, liable for a breach of Contract A.

The City of Nanaimo appealed. It argued that it was not applying undisclosed criteria but rather choosing the bid that it believed would give "the greatest value based on quality, service and price."

The Court of Appeal reviewed the case in light of the recent decision of the Supreme Court of Canada in the case of *M.J.B. Enterprises Ltd. v. Defence Construction (1951) Ltd.* The issue in that case was whether privilege clauses permitted the owner to award the contract to a non-compliant bidder. The court decided that it could not, but then went on to speak about privilege clauses generally.

The request for bids on the Hammond Bay contract included several such clauses:

> Article 18. Tender Rejection
>
> The Owner reserves the right to reject any or all tenders; the lowest will not necessarily be accepted.
>
> The City of Nanaimo reserves the right to waive informalities in or reject any or all tenders or accept the Tender deemed most favourable in the interests of the City of Nanaimo.

Article 19. Award

Awards shall be made on tenders that will give the greatest value based on quality, service and price. Preference shall be given to local suppliers where quality, service and price are equivalent.

In addition, the bid form submitted by Sound expressly provided that Nanaimo *"...is in no way obligated to accept this Tender."*

In *M.J.B.*, the Supreme Court adopted with approval the traditional views stated in **Goldsmith on Canadian Building Contracts**:

> The purpose of the [tender] system is to provide competition, and thereby to reduce costs, although it by no means follows that the lowest tender will necessarily result in the cheapest job. Many a "low" bidder has found that his prices have been too low and has ended up in financial difficulties, which have inevitably resulted in additional costs to the owner, whose right to recover them from the defaulting contractor is usually academic. Accordingly, the prudent owner will consider not only the amount of the bid, but also the experience and capability of the contractor, and whether the bid is realistic in the circumstances of the case. In order to eliminate unrealistic tenders, some public authorities and corporate owners require tenderers to be pre-qualified.

In other words, the decision to reject the "low" bid may in fact be governed by the consideration of factors that impact upon the ultimate cost of the project.

Writing for the Supreme Court, Justice Iacobucci said:

> Therefore even where, as in this case, almost nothing separates the tenderers except the different prices they submit, the rejection of the lowest bid would not imply that a tender could be accepted on the basis of some undisclosed criterion. The discretion to accept not necessarily the lowest bid, retained by the owner through the privilege clause, is a discretion to take a more **nuanced view of "cost"** than the prices quoted in the tenders.... [emphasis added]

Chief Justice McEachern, writing for the Court of Appeal, concluded that

> ... the privilege clauses in the request for tenders releases Nanaimo from the obligation to award the work to the lowest bidder if there are valid, objective reasons for concluding that better value may be obtained by accepting a higher bid.

The previous dealings between Nanaimo and Sound provided the basis for the additional criteria addressed by the owner's staff who, in the opinion of the court, have not been shown to have acted unfairly or other than in good faith in determining which bid provided the "greatest value based on quality, service and price" to the City. The court would not substitute its own analysis for that of the owner in whom the discretion to award the contract ultimately resided.

The consideration of past dealings between the parties did not constitute an undisclosed criterion:

In fact, past dealings are probably the best indicator of how a proposed relationship will come to work out in practice. I would caution, however, that this discretion must not be exercised in such a way as to punish or to get even for past differences.

Justice McEachern concluded that Nanaimo had shown sufficiently good reason for its decision to award the construction contract to the second lowest bidder. In a unanimous decision, the court allowed the appeal and dismissed Sound's action.

When the Court of Appeal reversed Madam Justice Downs, it used a more nuanced view of "cost" — in the form of claim history — to absolve the owner from a breach of Contract A. The appellate court reasoned that the trial Judge did not give sufficient weight to a provision in the bid documents which conferred flexibility on the owner.

The British Columbia Court of Appeal appears to have expanded the field of nuance identified by the Supreme Court of Canada in *M.J.B.* Claim history was not enunciated by the Supreme Court of Canada but was blessed by the British Columbia Court of Appeal.

Mr. Justice McEachern acknowledged that the he found the expansion of the nuance family "worrisome" because it created an opportunity for "arbitrariness". Some would argue that arbitrariness has no place within the embrace of an implied duty of fairness.

In *Sound Contracting*, the trial Judge held that claims history was an undisclosed criterion — a bit like the local preference in *Chinook Aggregates*. Madam Justice Downs also held that, having developed an undisclosed criterion, the owner then applied it unevenly. The British Columbia Court of Appeal did not address unevenness. One might conclude that it is fair for an owner to apply a Contract A evaluation criterion against one bidder while refraining from applying the same criterion to another. This seems incompatible with the notion of treating all bidders fairly — meaning that the same yardstick is used the same way on all bidders.

## MIDWEST MANAGEMENT (1987) LTD. v. B.C. GAS UTILITY LTD.

Court of Appeal for British Columbia; October 2000

After almost 20 years of litigating bidding issues, the construction and legal industries could be forgiven for thinking that all the possible angles on the *Ron Engineering* decision have been explored. In particular, the so-called privilege clause has been debated *ad nauseam*. Even the Supreme Court of Canada had its say on the subject in *M.J.B.* Yet, the following bidding case manages to offer a completely new take on the privilege clause!

In addition, the case turned out to be an indication of battles to come regarding the duty of fairness — outside Contract A.

In 1989, B.C. Gas Utility Ltd. invited six general contractors to submit bids for the construction of an underground pipeline in British Columbia. One of the bidders was the joint venture of Midwest Management (1987) Ltd. and Monad Contractors Ltd., which in our summary will be called Midwest.

The Instructions to Tenderers issued by B.C. Gas contained the following provisions:

> The Tender Documents are to be completed exactly as requested in order that bids may be compared on a uniform basis... .
>
> Tenderers shall make no changes to the Tender Documents in format or text in any manner. If Tenderer wishes to take exceptions, it may do so in Statement of Discrepancies and Omissions, stating clearly whether an exception is a condition precedent to its Tender or at OWNER'S option....
>
> Transmittal letters from Tenderers which contain qualifications to their Tender shall be deemed not to be part of or to supersede the Tender, and may be disregarded by the OWNER....

The Detailed Specifications included in the bid documents contained the following passage with respect to subsurface conditions:

> The CONTRACTOR shall be solely responsible for conducting its own tests or investigations to satisfy itself as to the soil structure, groundwater, sub-surface conditions, ditch wall stability, obstructions and all other conditions it may expect to encounter during the course of carrying out the Project and shall include contingencies for changes and variations in the aforesaid in its unit prices.

Midwest's bid included a covering letter. It stated that Midwest had not provided any amount for dewatering costs and proposed that such work be reimbursed on a cost-plus basis. Midwest did not set this out in a Statement of Discrepancies and Omissions as required by the Instructions to Tenderers.

B.C. Gas reviewed the bids and eliminated three contractors. To each of the remaining contractors, it sent a Request for Clarification asking them to address specific faults in their tenders. At a meeting with Midwest's representatives, B.C. Gas pointed out to Midwest that its position on dewatering did not comply with the instructions. Midwest stuck to its guns.

An agreement on the issue of dewatering was never reached and, in the end, B.C. Gas sent a letter to Midwest advising that it would not be awarded the contract.

Midwest sued B.C. Gas for breach of contract (Contract A, as per *Ron Engineering*) and breach of a duty of fairness, which, Midwest alleged, B.C. Gas owed to all bidders.

B.C. Gas applied to the court for judgment dismissing Midwest's action. The application was heard in May 1999 by Justice Paris, in chambers. The following is a summary of his decision.

B.C. Gas's position was that Midwest's tender did not comply with the bid documents (which was not disputed by Midwest), so no Contract A came into existence. Therefore, the claim based on breach of contract must be dismissed.

Midwest's principal argument regarding the existence of a Contract A in this case was based primarily on the privilege clause. In its early years, this clause used to be short and to the point: "Lowest or any tender not necessarily accepted". In this case, we see it fully grown up:

> OWNER reserves the right to reject any or all Tenders, including without limitation the lowest Tender, and to award the Contract to whomever OWNER in its sole and absolute discretion deems appropriate notwithstanding any custom of the trade to the contrary nor anything contained in the Contract Documents or herein. OWNER shall not, under any circumstances, be responsible for any costs incurred by the Tenderer in the preparing of its Tender.
>
> Without limiting the generality of the foregoing, OWNER reserves the right, in its sole and absolute discretion, to accept or reject any Tender which in the view of the OWNER is incomplete, obscure, or irregular, which has erasures or corrections in the documents, which contains exceptions and variations, which omits one or more prices, which contains prices the OWNER considers unbalanced, or which is accompanied by a Bid Bond or Consent of Surety issued by a surety not acceptable to the OWNER.
>
> Criteria which may be used by the OWNER in evaluating tenders and awarding the contract are in the OWNER's sole and absolute discretion and, without limiting the generality of the foregoing, may include one or more of: price; total cost to OWNER; the amount of B.C. content; the amount of Canadian content; reputation; claims history of Tenderer; qualifications and experience of the Tenderer and its personnel; quality of services and personnel proposed by the Tenderer; ability of the Tenderer to ensure continuous availability of qualified and experienced personnel; the Construction Schedule and Plan; the proposed Labour and Equipment; and the proposed Supervisory Staff.
>
> Should the OWNER not receive any tender satisfactory to the OWNER in its sole and absolute discretion, the OWNER reserves the right to re-tender the Project, or negotiate a contract for the whole or any part of the Project with any one or more persons whatsoever, including one or more of the Tenderers.

Midwest turned this around. It argued that because of the privilege clause, B.C. Gas was perfectly capable of accepting Midwest's offer in spite of its irregularity. In fact, none of the other bids, even the one which was eventually accepted, complied entirely with the bid documents as to the details or specifications contained in them.

Midwest found a sentence in the *Ron Engineering* decision that seemed to support this view:

> If the tenderer has committed an error in the calculation leading to the tender submitted with the tender deposit, and at least in those circumstances where at that moment the tender is capable of acceptance in law, the rights of the parties under contract A have thereupon crystallized.

The privilege clause, argued Midwest's counsel, made Midwest's bid "capable of acceptance" because this clause, in effect, made *any* proposal capable of being accepted. Therefore, the principle in the *Ron Engineering* case quoted above applied, and Contract A "crystallized" when Midwest submitted its bid.

Counsel further maintained that a term must be implied as a matter of law into this Contract A — an implied duty of fairness that the owner owes to all bidders.

Justice Paris reviewed the legal principles governing Contract A, and several binding decisions. In particular, the decision of the Supreme Court of Canada in the *M.J.B. Enterprises* case contradicted Midwest's novel argument that because of a privilege clause, any serious offer, even if non-compliant, brings into existence Contract A. There is no Contract A if the bid does not strictly comply with the requirements of the bid documents published by the owner. "That did not happen here and the train stops there", said Justice Paris. Midwest's bid was, at best, a counteroffer to B.C. Gas's offer as contained in its bid documents. Without a contract, there could be no implied duty of fairness.

This did not end the matter. Midwest's counsel also submitted that B.C. Gas's continued dealings with Midwest in requesting clarification and so on after the submission of the bid meant that, in effect, B.C. Gas treated Midwest's bid as valid. This, he said, brought Contract A into existence.

Justice Paris drew the opposite conclusion, namely, that the process of clarification indicated that B.C. Gas treated Midwest's tender as *not* valid because it did not comply with the bid documents. The owner simply wished to negotiate further with Midwest.

The court, therefore, dismissed Midwest's claim for damages for breach of contract.

However, that was still not the end of the case. Midwest further maintained that even if Contract A did not come into existence, the surrounding circumstances of the call for bids, the submission of bids and the subsequent dealings between the parties gave rise to a "free-standing" duty of fairness. A breach of that duty could result in liability independent of contract.

But what was the nature of the wrong alleged by Midwest? What legal pigeonhole could it be placed in? Midwest could only say that it might be a kind of tort.

There is no free-standing enforceable duty of fairness, said Justice Paris. It is hard to argue that a person does not have an obligation to act fairly towards others, or even that the law should not compensate a person who has suffered foreseeable damage at the hands of somebody who has knowingly acted unfairly towards that person. The problem lies in defining the content of such a principle which could be put into operation *as law*. Obviously, what is fair for one person is not necessarily fair for another, even when both are thinking and acting in perfectly good faith.

However, perhaps Midwest was right, continued the judge. The law might recognize a duty that would oblige an owner not to withhold information as to preconditions for acceptance of bids, if this could foreseeably cause harm to a prospective bidder. Such a duty may arise in the context of the tender process entirely apart from the existence of a Contract A.

This issue, whether it is called a duty of fairness or something else, could not be decided on a motion for dismissal without a trial. Therefore, since it was not evident that Midwest's claim was doomed to fail, the court dismissed B.C. Gas's motion for dismissal.

Midwest appealed. Justice Finch of the Court of Appeal for British Columbia agreed with the trial judge regarding Contract A:

> … [Midwest's] counter-offer for de-watering on a cost plus basis in its covering letter did not conform to the requirements of the tender documents, and [B.C. Gas] was fully justified in disregarding that proposal when it considered [Midwest's] bid.

He then turned his attention to the independent duty of fairness. Whether such a duty exists is a pure question of law, he said. The trial judge knew of no such duty. Counsel did not refer the court to any authority where such a duty has been held to exist. *"Such a duty is quite inconsistent with an adversarial, competitive tendering process. To find such a duty would cause great uncertainty in this area of the law,"* concluded Justice Finch.

As no such duty exists in law, he added, the claim based on its alleged breach was bound to fail.

Duties between parties — outside of a contract setting — fall into the family of tort. The British Columbia Court of Appeal dealt with the notion that a duty of fairness arose outside of Contract A by ruling that no such duty existed in law. In considering this result, the emphasis should fall on the word "fairness" rather than the word "duty".

As individuals, we owe others who share our space a variety of duties. Most of those duties involve taking care not to injure persons or property. And, in

CHAPTER 3: DUTIES AND RIGHTS OF OWNERS  57

some narrow cases, a duty not to cause pure economic loss — a loss not connected to property damage or personal injury.

There are several cases discussed in this book where a duty of care — of one kind or another — was found in a bid setting where Contract A had not come into being and where the duty was alleged concurrently with a contract obligation under Contract A. In none of those cases has a court recognized a duty of care expressed as a duty of fairness.

In *Martel*, summarized below, a duty of care in negotiations — expressed as a duty to share competitive information — was defeated by a policy decision of the Supreme Court of Canada. *Martel* was a pre-bid negotiation involving pure economic loss. If a duty of fairness existed at law — and currently it appears that it does not — then damages for its breach would result in compensation for pure economic loss. Since the Supreme Court of Canada policy decision in *Martel* was also dealing with pure economic loss, the same fate is likely to befall a free-standing duty of fairness.

So it seems that a free-standing duty of fairness has two formidable hurdles to clear. First, the British Columbia Court of Appeal says it does not exist in law. Second, the Supreme Court of Canada has held — as a matter of policy — that it will not "hobble" commercial negotiations by imposing duties between the parties which, if breached, would bring compensation for pure economic loss.

Don't bother knocking on the door marked "free-standing duty of fairness". No one is home — for the moment.

## MARTEL BUILDING LTD. v. CANADA

Supreme Court of Canada; February and November 2000

Martel Building Ltd. was worried. Its ten-year lease to a Crown agency would expire August 31, 1993. Rental rates in Ottawa were falling. Martel hoped it could negotiate a renewal and avoid a bid process by the Crown.

In early 1991, Martel told the Crown that it wished to negotiate a renewal and that it planned to upgrade the building. The Crown was interested. The individual delegated to negotiate with Martel did nothing. Nevertheless, he advised his superior that a proposal was coming from Martel.

Feeling stymied, Martel initiated negotiations in April of 1992. It was well-known that, later that year, bids would be called absent a renewal.

Martel failed to meet the Crown's rental targets. Preparation of bid documents began. Meetings between Martel and the Crown continued as the deadline approached. When Martel offered the rent the Crown wanted, price in hand, the Crown demanded — for same-day delivery — full details of Martel's planned upgrade. Otherwise, the bid process would proceed. Martel could not deliver and bids were called.

The bid documents gave the Crown wide berth on bid evaluation. In addition to the privilege clause (*"the Lessor may accept any Offer whether it is the lowest or not or may reject any or all Offers"*) the Crown reserved the right to consider, besides rent, its tenant costs (fit-up), and other expenses.

Martel submitted the low bid. After analysis, the Crown added $1,000,000 to Martel's price for fit-up and a further $60,000 for a security card system. The Crown performed the same analysis and added fit-up to the other bids. But, those bids were not saddled with the cost of the security system.

Now the second lowest bidder, Martel lost its tenant.

Martel sued the Crown for the lost rent, a "pure economic loss" (that is, financial loss without personal injury or property damage), claiming:

- breach of a duty of care in pre-bid negotiations;
- breach of a duty of care to consider Martel's interests in the preparation of the bid documents.
- breach of a contract obligation to treat Martel fairly;
- as an alternative to breach of contract, breach of a duty of care to treat Martel fairly.

Martel lost in the Federal Court Trial Division.

In the Federal Court of Appeal, Martel was successful. The Crown obtained leave to appeal to the Supreme Court of Canada.

The Supreme Court of Canada held that the landlord/tenant relationship created a duty of care which the Crown breached by negotiating negligently. Usually, after such a finding, the wronged party wins if it proves that the breach caused its damage — for Martel, lost rent. That prospect raised difficult policy questions for the Supreme Court. What standards of behaviour should be imposed on parties to commercial negotiations? How far should the law go to compensate a party which has suffered pure economic loss?

The Supreme Court of Canada considered five different policy scenarios and concluded that it would neither recognize the duty of care nor compensate Martel's type of pure economic loss. One of the five scenarios describes a negotiation model which transcends Martel:

> It would defeat the essence of negotiation and hobble the marketplace to extend a duty of care to the conduct of negotiations, and to label a party's failure to disclose its bottom line, its motives or its final position as negligent. Such a conclusion would of necessity force the disclosure of privately acquired information and the dissipation of any competitive advantage derived from it, all of which is incompatible with the activity of negotiating and bargaining.

In the lower courts, Martel claimed that the Crown failed to bargain in good faith. Martel did not argue good faith in the Supreme Court of Canada but that Court commented anyway:

> As a final note, we recognize that Martel's claim resembles the assertion of a duty to bargain in good faith. The breach of such a duty was alleged in the Federal Court, but not before this Court. As noted by the courts below, a duty to bargain in good faith has not been recognized to date in Canadian law. These reasons are restricted to whether or not the tort of negligence should be extended to include negotiation. Whether or not negotiations are to be governed by a duty of good faith is a question for another time.

The bid component of Martel's claim was one part contract and two parts tort.

Martel claimed that the Crown had breached Contract A which is created when a compliant bid is submitted in response to a bid call.

One tort component was the Martel claim of an independent duty to treat all bidders fairly — outside Contract A.

Martel also claimed that the Crown owed it a duty to prepare the bid documents carefully — meaning, with due regard to Martel's interests.

For over a decade, Canadian appellate courts have held that Contract A includes an implied duty to treat all bidders fairly — unless the bid documents say otherwise. Until *Martel*, the Supreme Court of Canada had not expressly endorsed that view.

The Supreme Court of Canada examined the evaluation provisions of the bid documents and the privilege clause. It concluded that the implied obligation of fairness was part of this particular Contract A.

Then, the Supreme Court of Canada looked at how the Crown had applied the evaluation criteria and whether it satisfied the threshold of fairness. Two areas received special attention.

Martel claimed that the addition of $1,000,000 in fit-up costs was unfair. Not so, said the Court. The bid documents allowed it. The Crown used the same means of calculation and applied the outcome to all bidders.

The Court then looked at the $60,000 imposed on the Martel bid for the security system and found no reference to the system in the bid documents. It also looked to see if the system cost was added to all bidders — it wasn't.

The Court concluded that the Crown had committed two sins. First, it added a $60,000 item not disclosed in the bid documents (unfair). Second, it added this item to Martel's bid only (unequal).

The Supreme Court of Canada then examined the practical implications of the Crown's action. Although the Crown had been unfair, it reasoned, removal of $60,000 from the Martel total still left it second.

There was a foul. But, there was no harm. The Court dismissed Martel's claim for a breach of Contract A.

In approving the implied duty of fairness, Supreme Court of Canada included the following statement:

> While the Lease Tender Document affords the Department wide discretion, this discretion must nevertheless be qualified to the extent that all bidders must be treated equally and fairly. Neither the privilege clause nor the other terms of Contract A nullify this duty. As explained above, such an implied contractual duty is necessary to promote and protect the integrity of the tender system.

The Supreme Court of Canada then reviewed the alternative claim that the Crown breached a duty of fairness outside Contract A. It dismissed this tort claim with the following statement:

> In our view, the enumeration of the alleged foregoing breaches clearly reveals that the contract analysis, as canvassed above, subsumes any duty of care that Martel seeks to have recognized under tort. In this connection, we acknowledge that it is well established that an action in tort may lie notwithstanding the existence of a contract. However, it is equally clear that in assessing whether a tortious duty should be recognized where a contract already defines the rights and obligations of the parties in a chosen relationship, courts will look to the contract as informing that duty....

> However, in the circumstances of this case, regardless of whether there exists a co-extensive duty in tort to treat tenderers fairly and equally in evaluating the bids, Martel's tort claim cannot succeed for the same reasons that a contractual claim would fail. The duty of care alleged in tort in the case at bar is the same as the duty which is implied as a term of Contract A....

The Supreme Court of Canada rejected the claim that the Crown had to consider its past relationship with Martel in drafting the bid documents.

> First and foremost, we agree with the Department that it would call into question the integrity of the tender process if, by reason of a past relationship with, or special knowledge of, a potential bidder, there could be an enforceable obligation to take the interests of that particular bidder into account.

. . .

> To recognize that the Department owed a duty to Martel would be inconsistent with the basic rationale of tendering .... In this respect, it is imperative that all bidders be treated on an equal footing, and that no bidder be provided differential treatment on the basis of some previous relationship with the party making the call for tenders. It would defeat the purpose of fair competition to allow one bidder to be given some advantage from its previous dealings....

Tort, contract, concurrent liability, damages, public policy: in *Martel*, the Supreme Court of Canada left nothing to the imagination.

*The debate is over: an implied duty of fairness is a term of Contract A — unless the bid documents exclude it.*

The non-Contract A branches of the Martel claim also find purchase in a bid situation — or in circumstances which often flow from a bid process. One branch involves the obligations of the parties during negotiation. The other addresses the existence — or not — of a free-standing duty of fairness.

The policy decision made by the Supreme Court of Canada around negotiation confirms what most negotiators always felt — it is a bare knuckle affair. According to Martel, a party is entitled to keep to itself knowledge or information which confers a competitive advantage. There is no duty to disclose. The Court did not endorse fraud or misrepresentation as an acceptable negotiating tool. Otherwise — as Cole Porter poetically observed — "anything goes".

The Court dismissed the notion that the Crown had an independent duty of fairness outside Contract A as created by the Crown's bid process. In Martel, the Court held that a duty of care in tort (if there was one) was identical to the one created by Contract A. So, the Court reasoned it was not necessary to deal with the free-standing variety. This leaves open the question apparently answered by the British Columbia Court of Appeal in Midwest Management. There, the appellate court held there was no free-standing duty of fairness under circumstances where Contract A had **not** arisen due to the non-compliance of Midwest's bid.

So, is there a freestanding duty of fairness independent of Contract? Probably not. The British Columbia Court of Appeal would not recognize such a duty — declining it emphatically. In Martel, when the Court addressed a duty of care between parties to a commercial negotiation, it overrode any duty of care with a policy decision. So, it would appear that a freestanding duty of fairness has an uphill struggle for survival.

---

*The central question in the following appeal decision was whether the request for proposals (RFP) issued by the owner constituted a bid document intended to create a binding Contract A between the owner and the successful proponent, or whether the process was simply a "beauty contest," intended merely to be a non-binding invitation to enter into negotiations.*

## CABLE ASSEMBLY SYSTEMS LTD. v. DUFFERIN-PEEL ROMAN CATHOLIC SCHOOL BOARD

Court of Appeal for Ontario; February 2002

In 1995, the Dufferin-Peel Roman Catholic Separate School Board issued a Request for Proposals for a computer cabling project for its elementary schools.

The Board's RFP anticipated that the actual installation of the cabling would be broken into two phases. Slightly less than one-half of the work would be completed in the Board's 1995 calendar/budget year, and the remainder in the following year, "pending budget approval" in that year.

The RFP also set out the Board's objectives, concluding with the following remarks:

> The [Board] requires solutions covering *all* portions of this RFP. **Partial bids will not be accepted.** Please recall that the board will require the installation of approximately 40 sites in the 1995 budget year, and hopes to receive allocations to complete installations in the 1996 budget year. [Emphasis in original.]

The RFP also contained two privilege clauses, the first of which read:

> The Board shall not be obligated to accept the lowest, or any, proposal in whole or in part thereof and reserves the right to re-bid or cancel the project in its entirety.

The second privilege clause said the same thing in slightly different words. Finally, the RFP provided that the Board's Purchasing Acquisition and Disposal Procedures would apply. Under the definition section of the Procedures, "proposal" was defined as

> ... an offer from a supplier to provide goods and services, acceptance of which may be subject to further negotiation.

The Procedures also provided for "purchase by negotiation" to apply in various circumstances including, under clause 4.4, where *"the lowest bid received substantially exceeds the estimated cost of the goods."*

The evidence at trial disclosed that, at the time the RFP was made, the estimated cost of the goods or budget allocation for the entire project was $2,000,000.

Cable Assembly Systems Ltd. submitted the lowest acceptable bid, and the only one that came below the estimated cost for the project, at $1,914,430. However, by the time the proposals were evaluated, the Board was aware that it did not have the funding it had hoped to receive for 1996. The budget allocation was therefore down to $1,000,000.

Given the time constraints which did not reasonably allow for the issuing of a new RFP, and the fact that the lowest acceptable bid exceeded the revised "estimated cost of the goods," the Board decided to invoke clause 4.4 of the

Procedures and to proceed with purchase by negotiation. The Board met and entered into negotiations with each of the three lowest acceptable bidders, including Cable and Compucentre Toronto Inc. Each was given approximately 24 hours to submit a further bid, and did so.

Compucentre was asked for further clarification of its bid, and provided it. After clarification, its bid was $1,156,887 compared with Cable's new bid of $1,339,020. The Board issued a purchase order to Compucentre, and the project was completed.

Cable sued the Board for breach of the duty of fairness and good faith.

Before the trial, the parties agreed that the Board's RFP gave rise to a duty to act fairly and in good faith during the contract award process. The amount of damages was also agreed, so that the sole issue at trial was whether the Board had breached its duty.

At trial, it was Cable's position that

(a) the Board had acted unfairly in negotiating the terms of the final contract with the three lowest acceptable bidders rather than with Cable alone, as the lowest of the three.
(b) alternatively, that the Board breached its duty by failing to provide the three bidders with an equal opportunity to make a new proposal, and by treating them differently.

The trial judge, Justice Kent, reviewed the law relating to bidding, and found that the Board had a duty to be fair to all bidders and to act in good faith. He compared the recent decision of the Supreme Court of Canada in *M.J.B. Enterprises Ltd. v. Defence Construction (1951) Ltd.* to the classic 1981 decision of the same court in *R. v. Ron Engineering & Construction (Eastern) Ltd.*, and found that the *M.J.B.* decision allowed a more subjective analysis of the evidence to determine the intentions and reasonable expectations of the parties, whereas *Ron Engineering* had applied "*a more formalistic or technical approach.*"

He found no case law supporting Cable's claim that the Board was, at least initially, obliged to negotiate only with Cable. However, the cases made it clear that a purchaser cannot normally reject the lowest bid without legitimate cause, even if it has reserved the right not to accept the lowest bid. Where bidders are asked for cost reduction proposals, the purchaser is not permitted to bid shop. The purchaser must be fair to all bidders, avoid mischief and give no unfair advantage to any bidder.

The trial judge concluded with a rule: "*The bidders are entitled to know that both they and the purchasers are bargaining in good faith.*"

The RFP left alive the possibility of negotiation. Therefore, said the judge:

> ... it could not be said that either Cable or the Board intended that the submission of a proposal or bid would constitute either the initiation of contractual relations or the creation of contractual obligations. The circumstances were not such that Cable was entitled to more than serious consideration of its bid and fair treatment in subsequent negotiations.

But did the Board breach its duty of fairness? Was it fair to meet and negotiate with two other bidders in addition to the low bidder?

While a public board has a duty to its bidders to conduct itself in a manner that maintains the integrity of the bidding process, said the judge, it has the additional duty to taxpayers to act in a financially responsible manner. He continued:

> While a safer course for the Board may have been to negotiate only with the lowest bidder, it was not excluded from wider negotiations by the terms of the request for proposal, its own purchasing acquisition procedures or current jurisprudence.

He therefore found no breach of duty on this ground and dismissed both claims. Cable appealed, and, almost exactly two years later, the trial decision was reviewed by the Court of Appeal.

In the Court of Appeal, Cable argued that the Procedures did not allow for such negotiations to take place. It argued that the words "estimated cost of the goods" in clause 4.4 could only refer to the Board's initial estimate of $2,000,000, and that since Cable's bid did not exceed that amount, the clause did not apply. Cable therefore argued that, in the circumstances, the Board's only alternative was to reject all the proposals and to make a new RFP.

In the appellate court's view, based on the terms of the RFP and the Procedures, the trial judge's conclusion that wider negotiations were open to the Board was reasonable. By its terms, the RFP made it clear that the scope of the project and, consequently, the "estimated cost of the goods" were subject to funding being available for the year 1996. The court therefore rejected this ground of appeal.

Did all three bidders have an opportunity to make a new bid? The trial judge was satisfied that the Board understood the obligation to act fairly and in good faith. Its staff knew that one bidder's price could not be revealed to another bidder.

Cable contended at trial that the Board had not provided all three bidders with an equal opportunity to make a new bid because Compucentre, according to Cable, had received confidential information about the Board's revised budgetary position early, and had arrived at the first meeting with the Board prepared to discuss a revised proposal that involved fitting all the sites with half the cables as a means of lowering the project's cost, a suggestion that coincided with the Board's revised plan.

The trial judge was not satisfied on the evidence that there had been a leak of information. He held that Compucentre could have thought of the Board's plan independently. He was unable to find it more probable than not that such a leak had occurred; he therefore rejected the claim that the Board could be liable on that basis.

Given the express references in the RFP to the uncertainty of funding for the second year, the Court of Appeal was also of the view that the trial judge's finding on this point was supported by the evidence.

At trial, Cable also submitted that the Board accorded different treatment to the three bidders during the negotiating process, to the advantage of Compucentre. The trial judge held that there had been "differing treatment" in certain respects, and found that this was "unfortunate and inappropriate" but that it did not enable Compucentre's new bid to become lower than Cable's. Therefore, the question whether this constituted unfairness was "of academic interest only."

The trial judge's conclusion that the conduct in question had no bearing on the ultimate award of the contract was also supported by the evidence, decided the Court of Appeal. Consequently, since Cable did not lose the opportunity to participate fairly in the process, its claim was properly dismissed.

In the result, the Court of Appeal unanimously dismissed Cable's appeal, with costs.

Some purchasing authorities continue to believe that names rule. So, if you call a bid process an RFP, Contract A — under the *Ron Engineering* case law — will be avoided.

Owners should remember the "duck principle": if it looks like, sounds like, walks like and otherwise behaves like a duck, then it is a duck. In this case, a duck was a duck and Contract A was ruled to exist between the Board and at least three other bidders.

Both the trial court and the Court of Appeal ruled that the Board discharged its obligation under Contract A to treat all bidders fairly. This although the Board convened a three-way auction on an amended work scope and treated the three candidates differently.

The moral of the story is that the courts — and not authors of books — will have the last word on what does, and does not, discharge the Contract A duty of fairness.

---

*Following the decision of the Supreme Court of Canada in the M.J.B. case, compliance with the requirements of the bid documents became a "hot topic" in many bidding disputes. Bidders who failed the compliance test lost a chance to win the construction contract; owners who awarded the contract to a non-*

compliant bidder often ended up paying large sums of money to aggrieved bidders who, but for this breach, would have got the contract.

But what criteria should be used in assessing compliance? What kind of test? Does the bidder have to comply strictly with the bid requirements, or is substantial compliance sufficient?

The following case contains a useful discussion of this issue.

## J. OVIATT CONTRACTING LTD. v. KITIMAT GENERAL HOSPITAL SOCIETY

Court of Appeal for British Columbia; May 2002

In 1998, J. Oviatt Contracting Ltd. submitted a bid for site preparation work put out to bid by the Kitimat General Hospital Society in connection with the construction of the local Community Health Centre in Kitimat, B.C.

The work put out to bid involved stripping soil, stock piling it on and off site, installation of water diversion features and placement and compaction of imported granular fill. Oviatt was the low bidder at $583,620, but the work was awarded to the second lowest bidder, Boden Construction Ltd. Boden's bid was $596,211.

Oviatt sued the Hospital for damages for breach of contract and breach of the Hospital's duty of fairness for not awarding the contract to Oviatt.

At trial, Justice Hunter reviewed the events leading up to the contract award. The Hospital acted throughout on the advice of its geotechnical consultant, Agra Earth and Environmental Ltd. An Agra engineer, Jay McIntyre, reviewed the bids and recommended that the Boden bid be accepted.

Richard Grimsdell, an architect employed by the B.C. Ministry of Finance and a member of the Provincial Project Building Committee for the project, asked McIntyre for more detail in support of Agra's recommendation. McIntyre, in a letter to the Hospital copied to Grimsdell, gave three reasons:

- Boden's unit price for granular fill was lower, $9.53 compared to Oviatt's price of $11.14;
- Oviatt's bid left out four pages of the bid form;
- Oviatt qualified its bid by excluding a temporary road.

Grimsdell reviewed the bids and also recommended acceptance of the Boden bid. The price differential between the Boden and Oviatt bids for the granular fill was the main factor influencing the Agra recommendation. Agra recognized that the elevation of slabs for the buildings might have to be raised and more granular fill required than estimated in the bid. Selecting Boden minimized the risk of increased costs for more fill if the slab elevations were raised.

The decision is of particular interest because the trial judge discussed in some detail the test to be applied to determine whether Oviatt's bid was compliant or not. Was strict compliance with bid documents required, or only substantial compliance? The Supreme Court of Canada did not address this issue in *M.J.B.* because there was no dispute regarding the trial judge's finding that the bid was non-compliant.

Justice Hunter concluded that the appropriate test was one of *substantial compliance*. He reviewed some of the precedents.

In the grandfather of the bidding cases, *Ron Engineering*, the issue was already addressed by Justice Estey who said:

> It would be anomalous indeed if the march forward to a construction contract could be halted by a simple omission to insert in the appropriate blank in the contract the numbers of weeks already specified by the contractor in its tender.... It would be otherwise, of course, if a material fact were omitted from the tender, or if the meaning of the tender was unclear....

In the 1993 decision of the B.C. Court of Appeal in *British Columbia v. SCI Engineers & Constructors Inc.*, there were specific instructions that any corrections by fax must reveal only the amount by which the bid was being altered, not the revised total. Prior to the opening of bids, SCI, one of the bidders, nevertheless stated the revised total in its faxed revision. Chief Justice McEachern made the following statement:

> ... we think there are no circumstances in this case which require the Crown to apply a strict rather than a substantial compliance test, particularly when the Crown was satisfied that no confusion was caused by the last revision.

In *Foundation Building West Inc. v. Vancouver (City)*, the bid documents issued by the City specified that any changes to the bids were to be made in writing. One bidder telephoned the City ten minutes before the deadline and said that while the subtotals in its bid were correct, it had made a mathematical error in adding up the total. The City corrected the error and this resulted in that bidder having the lowest bid. Justice Sigurdson stated:

> There is no requirement in the tender documents here that the City reject a bid because it does not comply strictly with the tender documents. If the oral communication of the fact that there is an addition error amounts to a revision and is not in compliance with the tender documents, the City is not obliged in this case to reject it. .... To my mind, the circumstances and the tender documents in this case call for the application of the substantial compliance test as described in SCI, not a strict compliance test.

The judge also found in the Hospital's bid documents provisions which apparently supported the application of the substantial compliance test. For example, the Hospital reserved the right to "waive irregularities in the bid form" if it felt that such irregularities were of "a minor or technical nature". On the other hand, there was no provision for the Hospital to reject a bid if it did not comply strictly with those documents.

Accordingly, Justice Hunter found that it was appropriate to apply the substantial compliance test to determine if Oviatt's bid was in compliance with the bid documents.

In the end, the judge concluded that Oviatt's action failed on two grounds. First, Oviatt's bid was qualified because it did not include a temporary road required by the bid documents. Without the road, Oviatt's bid was non-compliant even on the basis of a substantial compliance test. Second, he decided that, even if Oviatt's bid *was* compliant, the Hospital had not acted unfairly or in bad faith in accepting Boden's bid instead of Oviatt's.

Oviatt appealed the trial decision but lost in the Court of Appeal.

---

*This decision more firmly established in the firmament that, unless the bid documents say otherwise, strict compliance is not required — substantial compliance will do. This is sensible in every way and consistent with practice before Ron Engineering.*

*Where Oviatt did not make a contribution to the refinement of Ron Engineering case law was in the trial judge's determination of when Contract A came into being. Mr. Justice Hunter held that the sequence of analysis was to first determine whether Contract A existed and then to determine whether or not the bid was compliant. He put the cart before the horse, so to speak.*

*When it affirmed the trial decision, the Court of Appeal left Mr. Justice Hunter's analysis alone. It is tempting to think that it doesn't matter. After all, the key holding in the case involved the theory of substantial performance. But, the Supreme Court of Canada judgment in M.J.B. — not to mention Ron Engineering — made it clear that Contract A could only come into being with a bidder who submitted a compliant bid. First the horse, then the cart.*

## **WIND POWER INC. v. SASKATCHEWAN POWER CORP.**

Saskatchewan Court of Appeal; May 2002

This appeal concerns the authority of a crown corporation to reject all proposals after concluding that one of them met its RFP requirements.

Saskatchewan Power Corp. (SaskPower) requested proposals for a wind power project. It determined that the proposal of Wind Power Inc. came within its estimated costs and, accordingly, SaskPower's Board of Directors approved it.

SaskPower did not advise Wind Power that the Board had accepted their bid, but sought the approval of the Lieutenant Governor in Council as it was obliged to do under Saskatchewan's *Power Corporation Act* (PCA).

Cabinet ultimately directed SaskPower not to proceed with the project. Citing economic considerations, SaskPower advised the proponents that the project was cancelled. Wind Power sued.

SaskPower's RFP did not share all the characteristics of a bid but SaskPower accepted that the law of bidding may be applicable. Based on this acknowledgment, the trial judge applied the law as though the RFP was, indeed, a request for bids. The appeal was later argued on the same basis. The central issue was whether the privilege clauses in the bid documents allowed SaskPower to reject all proposals.

At trial, Wind Power's main arguments were that SaskPower:

- breached a term of Contract A requiring them to be awarded Contract B,
- had breached a duty of fairness owed to Wind Power, and
- had negligently misrepresented that SaskPower had the authority to proceed to Contract B.

SaskPower defended against these claims by relying on three privilege clauses, which reserved the right to reject any or all proposals, or to determine which proposal, if any, would be selected. It also argued that the PCA required that the approval of the Lieutenant Governor in Council had to be obtained.

Relying on the privilege clauses and the PCA, SaskPower asserted that it had the right to reject all the proposals. The trial judge agreed. She said:

> In this case there was no obligation on [SaskPower] to accept any bid. It had a right to accept any particular bid, if it wished. The failure by [SaskPower] to exercise that right in regard of any particular bid does not give an unsuccessful bidder a cause of action.

The trial judge was not convinced either by Wind Power's allegation that SaskPower breached its duty of fairness. Wind Power asserted that it had no knowledge of government involvement in the affairs of SaskPower which was not disclosed. The judge found that Wind Power's actions in lobbying for the project indicated otherwise.

As to negligent misrepresentation, the trial judge found that SaskPower officials, government officials, and ministers of the Crown had indeed made representations that the project would proceed but that Wind Power had not relied on them.

Wind Power's appealed on many grounds. It submitted that:

- there was a breach of the terms of Contract A between Wind Power and SaskPower because SaskPower failed to treat all bidders fairly and used undisclosed criteria in evaluating the bid;
- the privilege clause is not a defence to a breach of Contract A;
- the PCA is not a defence to a breach of Contract A;
- SaskPower is liable to Wind Power for damages even if SaskPower did not have Cabinet approval to enter into Contract B;
- Contract B was formed between Wind Power and SaskPower at the moment SaskPower's Board of Directors accepted Wind Power's bid, even if the acceptance was not communicated to Wind Power;
- in any event there was a communication of the acceptance of Contract B by SaskPower to Wind Power;
- there was an implied term in Contract A that SaskPower would proceed with Contract B, if an acceptable proposal could be found, barring unforeseen circumstances, and that there were no unforeseen circumstances.

Did Contract A arise in this case? Justice Jackson, who wrote the decision of the Court of Appeal, was persuaded that it was the intention of the parties that Contract A would arise upon the submission of a proposal. The principal issue was the precise nature of SaskPower's contractual obligations under this Contract A.

An important factor in the appeal was that, following the *Wind Power* trial decision, the Supreme Court of Canada released its decisions in *M.J.B. Enterprises Ltd.* and *Martel*. Each of them implied a term in Contract A which lessened the effect of a privilege clause.

Wind Power claimed that, in keeping with these two decisions, the Court of Appeal should imply a term in Contract A that SaskPower would proceed to negotiate Contract B if a suitable bid was submitted, and assuming no unforeseeable circumstances had arisen. This would be consistent with the reasonable expectations of the parties, preserve the integrity of the bidding process, and constitute an implied term in every bidding contract.

The Court of Appeal found much that was of merit in Wind Power's position, but found that, on the whole, it did not withstand analysis.

The Supreme Court said in *M.J.B. Enterprises Ltd.* that the focus must be on the intentions of the actual parties, with emphasis on the word "actual." Justice Jackson found that, when SaskPower issued the RFP, it intended to enter into the construction contract but not at any cost.

The further difficulty was that the term Wind Power was asking the court to imply would either deny the existence of the PCA, or compel Cabinet to act in a particular way. The court was, in effect, being asked to imply a term which would compel SaskPower to enter into a contract and ignore the law.

Besides, what would the implied term contain? Wind Power was looking for one which would compel SaskPower to enter into Contract B if an acceptable bid were submitted to it, and barring unforeseen circumstances.

The difficulty with such a term would be, first, the notion of compelling someone to enter into a contract when the terms of such a contract would have to be negotiated. Then there was also the question of what would constitute unforeseen circumstances. It came as a surprise to SaskPower that Cabinet would not accept its recommendation — this was therefore an unforeseen circumstance both for Wind Power and for SaskPower.

The court concluded that it was not possible to imply a term in Contract A that SaskPower would enter into Contract B if an acceptable bid were presented to it.

Another significant argument raised by Wind Power was that SaskPower breached its duty of fairness by relying on hidden or undisclosed criteria in rejecting Wind Power's bid.

The duty of fairness, replied Justice Jackson, has never been construed as obligating the owner to accept any of the proposals where it chooses to reject all of them. Furthermore, it would offend the rule of law to imply a duty of fairness which would obviate the necessity of complying with the PCA. Furthermore, the PCA was part of the legislative basis of the Province. "To consider this to be hidden criteria stretches one's normal understanding of these words," said Justice Jackson.

Finally, Wind Power submitted that Contract B between SaskPower and Wind Power was formed at the moment SaskPower's Board of Directors resolved to approve Wind Power's proposal. That argument did not help either.

The RFP provided for a time lag of over two weeks between the selection of the successful proponent and the signing of the purchase contract. This indicated to the Court of Appeal that the parties did not intend to be bound by Contract B immediately upon its acceptance. That conclusion was confirmed by the provision in the RFP that *"SaskPower and the successful Proponent shall negotiate in good faith the final terms and conditions of the Power Purchase Contract."*

Finally, if Contract B arose at the moment when SaskPower's Board of Directors approved Wind Power's bid, the court would have to consider the validity of such a contract in view of the conflict between it and the legislative provisions of the PCA.

The Court of Appeal rejected Wind Power's appeal.

*Wind Power* gave the Saskatchewan Court of Appeal an opportunity to exercise both the *M.J.B.* and *Martel* decisions, not long after both were issued by the Supreme Court of Canada.

Wind Power argued unsuccessfully that the owner had an obligation to award it a contract in the face of more than one privilege clause included in the bid documents. The Court of Appeal refused to bypass the owner's intention which included the right to reject if it could not find a suitable contractor at a suitable price. Besides, forcing the owner to proceed meant implying a term into Contract A which would oblige the owner to ignore the law — the need to get Saskatchewan Cabinet approval.

Wind Power did not fare better when it argued a breach of the implied duty of fairness. Wind Power argued that the owner failed to disclose its statutory obligation to obtain cabinet approval — an undisclosed criterion. Not so, said the Court of Appeal. Wind Power was aware of the statutory framework. And the court was not prepared to enforce a duty of fairness which obliged the owner to break its own enabling law.

This is one of the few cases where the "pay off" for Contract A is the right to negotiate toward a Contract B. Wind Power attempted to argue that Contract B arose when the owner's Board passed a resolution approving Wind Power as the candidate for negotiation and recommending that Cabinet approve this decision. The Court of Appeal recognized this as a step in the process but it refused Wind Power's submission that Contract B arose at that moment.

Can Contract A take a successful bidder to something other than Contract B? Why not?

---

## MELLCO DEVELOPMENTS LTD. v. PORTAGE LA PRAIRIE (CITY)

Manitoba Court of Appeal; October 2002

In March 1998, the City of Portage la Prairie, Manitoba, issued a Request For Proposals (RFP) for the sale and development of certain City-owned land. The RFP identified "an exciting opportunity" for a residential development on a large piece of land. The document called for "concept plans for new residential areas" and stated that the City would negotiate with the applicant that presented the "most attractive proposal." It clearly stated:

> This is an invitation for proposals and not a tender call.

Only two proposals were received: a combined one from Lions Park Housing Inc. and Lions Club, and one from Mellco. Mellco made an unconditional offer to pay the proposed purchase price ($316,000) in full by August 7, 1998. Lions offered to pay $425,000 for the property.

In a number of respects the Lions' proposal deviated from the requirements of the RFP. Nevertheless, the City decided to accept the it as being "the most attractive to the City." It signed an agreement with Lions on June 15, 1998. Mellco sued.

The lengthy trial commenced on October 30, 2000, and was completed on December 3. In his reasons for decision delivered September 17, 2001, Justice Clearwater dismissed Mellco's claim with costs. The matter ended up in the Manitoba Court of Appeal.

Mellco's argument was that the City breached Contract A by considering the non-complying proposal from Lions at all, then compounded the error by failing to evaluate the proposals only in accordance with the RFP criteria. The City was bound to accept the best compliant proposal assessed under the RFP criteria alone.

Furthermore, Mellco submitted, the City was under a general duty to negotiate fairly in either a bidding or an RFP process.

The City responded that it was always its intention to maintain a complete and unfettered discretion to determine what was "most attractive" to it. At most, the RFP represented an opportunity for Mellco and others to submit a proposal which, if found to be "attractive," would give the proponents the ability to negotiate a development agreement.

The Court of Appeal agreed with the decision of the B.C. Supreme Court in *Powder Mountain Resorts Ltd. v. British Columbia*:

> The invitation for proposals appears to have been an invitation to negotiate or, in other words, an invitation to treat. It appears unlikely that the intention of the parties was that a submission of a proposal would initiate contractual relations between the parties. It appears more likely that the intention was to initiate negotiations which, if mutually satisfactory, would lead to contractual relations.

Canadian courts have consistently adopted the position that the law will not enforce an agreement to negotiate. Also, in the 1992 decision of the House of Lords in England in *Walford v. Miles,* Lord Ackner said:

> The reason why an agreement to negotiate, like an agreement to agree, is unenforceable, is simply because it lacks the necessary certainty.... How can a court be expected to decide whether, *subjectively*, a proper reason existed for the termination of negotiations? The answer suggested depends upon whether the negotiations have been determined "in good faith." However, the concept of a duty to carry on negotiations in good faith is inherently repugnant to the adversarial position of the parties when involved in negotiations.... A duty to negotiate in good faith is as unworkable in practice as it is inherently inconsistent with the position of a negotiating party .... Accordingly, a bare agreement to negotiate has no legal content.

Was the RFP a binding bid document obliging the City to enter into a development agreement with Mellco? The Court of Appeal relied on the normal principles of contractual intention to answer this question. In *M.J.B. Enterprises*, Justice Iacobucci of the Supreme Court of Canada summarized the principle in this way:

> ... whether or not Contract A arose depends upon whether the parties intended to initiate contractual relations by the submission of a bid in response to the invitation to tender.

For an owner, the point of issuing an RFP instead of a call for bids is that he wants submissions from interested parties but does not wish to create Contract A. Properly drafted, an RFP asks parties for expressions of interest and sets out the owner's intention to consider those expressions of interest, and then to undertake negotiations with one or more parties whose proposal(s) appeal to the owner.

The court concluded that the City's RFP was not intended to create a binding contractual relationship between the City and the "winning bidder." The fact that the proposal contained the statement *"This is an invitation for proposals and not a tender call"* was but one of the many factors militating against the applicability of *Ron Engineering* to the RFP process.

Where the final terms of the contract are contained in the bid (*i.e.*, there is no need for negotiation), courts will readily accept that the bid is not a mere "invitation to treat." However, it is not possible to identify in the RFP the terms of any Contract B. Fundamental details must first be finalized through discussions and negotiations. *"Cases such as this do not fall to be decided under the law of tenders as articulated in Ron Engineering,"* decided the Court of Appeal.

The opportunity to negotiate with the preferred proponent also precluded the need to imply a term to reject non-compliant bids, as Mellco maintained. In *M.J.B. Enterprises*, Justice Iacobucci stressed that *"the [owner] did not invite negotiations over the terms of either Contract A or Contract B."* Because of this, he said, *"it is reasonable to infer that the [owner] would only consider valid tenders."*

The City's RFP was meant to lead to negotiations. These gave the parties ample opportunity to resolve any difficulties inherent in the initial proposal so as to achieve the City's goal of accepting the "most attractive proposal."

Can the RFP process invoke the obligation of fair bargaining in good faith that is now firmly established in formal bidding cases?

The court agreed with Mellco that the question of the duty to negotiate in good faith with respect to bids or proposals, is a form of continuum. At one end are the formal bid cases based on the principles of *Ron Engineering*. At the other end are cases where, for example, an owner requests a simple quote. There is obviously a lot of territory between these two extremes.

The City was obliged to conduct itself fairly, and in good faith. Without some fairness in the system, proponents could incur significant expenses in preparing futile bids which could ultimately lead to a negation of the process. There must be enough fairness and equality in the procedures to ensure its integrity and openness.

The principles of fairness and good faith are not determined in a vacuum, said the court, but rather are implied, based on the intentions and expectations of the parties. In a *Ron Engineering* type of bidding process, the requirement of good faith and fairness is a term that is implied into Contract A. But there was no Contract A in a mere request for proposals opening up a process of negotiation.

The court was not at all persuaded that Mellco was treated unfairly, or that the City acted in bad faith. A purely objective evaluation in such circumstances was impossible. Furthermore, the process calling for "attractive proposals" followed by "negotiations" with the lead proponent stood in stark contrast to the formal bidding process in which bids are meticulously scrutinized for conformity. *"The duty to refrain from awarding a form of contract that is significantly different than Contract A has no application to the situation calling for proposals only,"* concluded the court and dismissed Mellco's appeal.

In *M.J.B.*, Mr. Justice Iacobucci underscored the critical underpinning of the existence of Contract A. In paragraph 19 of his judgment, he states:

> What is important, therefore, is that the submission of a tender in response to an invitation to tender may give rise to contractual obligations, ..., depending upon whether the parties intend to initiate contractual relations by the submission of a bid. If such a contract arises, its terms are governed by the terms and conditions of the tender call.

Courts are often confronted with procurement documents which are labeled something other than "bid". But, upon analysis, all of the elements which gave rise to Contract A in *Ron Engineering* turn out to be present. Call it a hippopotamus if you want. But, if the elements are present, a hippopotamus will be treated like a bid.

In *Mellco Developments*, the court did not find a process with an identity problem. The owner hung the "RFP" label on its process. And the process itself supported the conclusion that the owner was saying what it meant. Among other things, the owner included the following statement in the RFP: "This is an invitation for proposals and not a tender call."

The RFP went on to request "attractive proposals" and promised "negotiations" to those whose proposals found favour with the owner.

In *Mellco*, the Manitoba Court of Appeal made it clear that a true RFP — an invitation to negotiate — was not a vast wasteland free of duties owed by owners to RFP participants. Having invited parties to spend money making proposals, the owner had erected an RFP framework which imposed upon it obligations — duties to participants — which were consistent with the expectations which its process created. The appellate court found that the owner had not departed in any significant way from the process which it set in motion — meaning it met its non-contract A duty of care to follow its own process.

Contract A this was not.

In *M.J.B. Enterprises Ltd. v. Defence Construction (1951) Ltd.*, the Supreme Court of Canada decided that, as a term of Contract A, there was no obligation to award the contract to the lowest bidder. However, there was an obligation *not* to award the contract to a non-compliant bidder. A non-compliant bid does not represent acceptance of the owner's offer to enter into a Contract A. It is merely a counter-offer, and cannot create Contract A.

The decision has caused major headaches for owners.

They now have to double- and triple-check each bid for compliance because an error could cost them dearly — and still may end up paying the plaintiff bidder's lost profit if a judge disagrees with their assessment.

They have to reject an otherwise attractive bid maybe because the contractor, in the last-minute rush to submit the bid or for some other reason, had failed to comply with some requirement.

How does an owner respond to this challenge? By (a) simplifying the bid documents to eliminate non-essential requirements from the bidding process? Or, (b) by adding more legal language to the documents?

The answer, of course, is (b). The result is the so-called "discretion clause" in addition to (or combined with) the well-established "privilege clause." Where the latter reserves for the owner the privilege to accept or reject any bid, the first allows the owner to consider and accept, at its own discretion, a non-compliant bid.

## KINETIC CONSTRUCTION LTD. v. COMOX-STRATHCONA (REGIONAL DISTRICT)

British Columbia Supreme Court; November 2003

The facts of the *Kinetic* case are far from unusual. In June 2002, the Regional District of Comox-Strathcona in British Columbia issued a Request For Proposals for the construction of an infrastructure project. Six proposals were submitted; the lowest were those of Kinetic Construction Ltd. ($1,494,790) and D. Robinson Contracting Ltd. ($1,495,000).

A so-called "discretion clause" in the bid documents reserved for the owner the right to reject or retain bids *"which are nonconforming because they do not contain the content or form required by the Instructions to Tenderers,"* and even to use undisclosed evaluation criteria. It stressed that price was not the only or even primary criterion.

The bids were analyzed by an engineering consultant, Earth Tech Inc. The consultant preferred Robinson but found that its bid contained two qualifications which decreased the scope of the work covered by the bid. Nevertheless, in Earth Tech's opinion, the defects in Robinson's bid were minor, and the bid was acceptable. Based on the consultant's recommendation, the contract was awarded to Robinson.

Kinetic asked the B.C. Supreme Court to award it damages suffered as a result of not having been awarded the contract. Its main argument was that the Robinson bid was non-compliant and therefore could not be accepted.

Justice Preston (in chambers) reviewed the law of bidding.

A bidding contract, Contract A, is automatically created when a bidder submits a compliant bid, per *Ron Engineering*.

If there is a discretion clause in the bid documents allowing the acceptance of a non-compliant bid, then a non-compliant bid will not *automatically* give rise to a Contract A. However, said Justice Preston, there is no principle of law requiring the owner to reject a non-compliant bid if the bid documents expressly reserve to the owner the right to accept such bids. The discretion clause gives the owner the discretion to accept a non-compliant bid, and so enter into a Contract A with the bidder.

Kinetic's bid, which was compliant, created a Contract A with the District automatically, the moment it was submitted.

In contrast, Contract A arose between Robinson and the District when the District exercised its discretion to consider its bid despite the bid's non-compliance.

Once a non-compliant bid is admitted for consideration, a duty arises to treat the bidder fairly.

In *Martel Building Ltd. v. Canada*, the Supreme Court of Canada laid down the principle that fair and equal treatment of bidders is an implied term of Contract A:

> Without this implied term, tenderers, whose fate could be predetermined by some undisclosed standards, would either incur significant expenses in preparing futile bids or ultimately avoid participating in the tender process. A privilege clause reserving the right not to accept the lowest or any bids does not exclude the obligation to treat all bidders fairly.

"Nevertheless", added the court, "the bid documents must be examined closely to determine the extent of the obligation of fair and equal treatment."

Justice Preston discussed some other pertinent court decisions. In *Elite Bailiff Services Ltd. v. British Columbia*, the B.C. Court of Appeal examined the extent to which an owner must disclose the *weight* afforded to each item listed in the bid documents as an evaluation factor, and found that detailed disclosure of such criteria was not necessary.

In *British Columbia v. SCI Engineers & Constructors Inc.*, the B.C. Court of Appeal considered a bid which contained corrections that did not comply strictly with the conditions in the bid documents. The court found that the Province was under no duty to apply a *strict* rather than a *substantial* compliance test, particularly when it was satisfied that no confusion would be caused by the corrections. Chief Justice McEachern quoted from the *Ron Engineering* decision:

> It would be anomalous indeed if the march forward to a construction contract could be halted by a simple omission to insert in the appropriate blank in the contract the numbers of weeks already specified by the contractor in its tender.

The test of substantial compliance was also applied in *J. Oviatt Contracting Ltd. v. Kitimat General Hospital Society*, where Oviatt omitted four of eight pages of the bid forms, and did not fill in the blanks on pages requiring certain dates. The court found that Oviatt's omission of four pages of the bid form did not render its bid non-compliant. But, its proposal for a different scope of work did.

The Robinson bid clearly highlighted the two areas in which it did not comply with the Instructions to Tenderers. While the District may have been entitled to reject the Robinson bid for non-compliance, it chose not to do so. Justice Preston commented:

> ... owners awarding contracts are entitled, if they do so fairly, to consider bids that do not fully comply with the bid requirements that are imposed to achieve consistency. If they choose to proceed in this manner, they must disclose the nature and extent of their discretion in the information that they provide to tenderers and proceed in a manner that results in a fair comparison of the competing bids.

Kinetic's contention that the two qualifications in the Robinson bid decreased the scope of the work bid is "well-founded," added the judge. However, Earth Tech provided sufficient information to permit the District to fairly evaluate the magnitude of the difference between the bids. On the basis of this information, the District decided to award the contract to Robinson.

The District was entitled to make the decision it did on the basis of the bid process outlined in the Information to Tenderers, decided the judge, and dismissed Kinetic's claim.

What is troubling about the Motions Court decision in *Kinetic* is Mr. Justice Preston's sense that Contract A was created with the non-compliant bidder when the owner elected to retain and evaluate that bid. How can this be?

To create Contract A, there must be an offer from the owner (offer of Contract A in the form of an invitation to bid for Contract B), acceptance by the bidder (submission of a compliant bid) and consideration (offer for Contract B/Contract A evaluation by the owner).

In *Kinetic*, the non-compliant bidder rejected Contract A — by operation of law — when it made a counter offer for a different Contract B (the non-compliant bid).

There is nothing in contract law or in the bid documents which empowers the owner to create Contract A with the non-compliant bidder without going through the classic offer/acceptance/consideration mechanism. The bid documents speak only of an option to reject or retain the (non-compliant) offer.

Mr. Justice Preston imposes on the owner, having elected to "retain" the non-compliant bid, an obligation to treat that bidder fairly. For that to be so, the owner and the non-compliant bidder must have Contract A (let's call it Contract A+) and it must include an implied duty of fairness as no such duty is described in the bid documents.

The clause which permits the owner to retain the bid of the non-compliant bidder is a term of Contract A with the compliant bidders — one intended to protect the owner from suits by them. It is hard to imagine how the retention clause can create Contract A (or Contract A+) between the owner and the non-compliant bidder — at the unilateral election of the owner. This is important because if some version of Contract A does **not** exist, then *Midwest Management* and *Martel* have held that there is no free-standing duty of fairness or of good faith dealings owed by the owner to the bidder.

So, the discretion clause gives the owner the discretion to accept a non-compliant bid, and so enter into a Contract A with the bidder? Not according to the court in the *Graham* case, decided within a few days of the *Kinetic* decision by another judge of the same court, and supported by the B.C. Court of Appeal.

A summary of the *Graham* case appears in Chapter 6 "Mistake in Bid."

---

## SUMMARY

The following points can be distilled from the cases in this section:

- Contract A includes an implied duty of fairness unless excluded by the bid documents.
- The privilege clause will not protect an owner from a breach of Contract A.
- The privilege clause is (usually) incompatible with an award of Contract B to a non-compliant bidder.
- The privilege clause does not oblige an owner to award Contract B to a low compliant bidder.
- An owner will be held to an objective standard when it exercises its rights under a discretion clause.
- Whether a procurement process gives rise to Contract A or not depends on the intentions of the parties as reflected in the bid documents issued.
- Outside Contract A, there is no free-standing duty of fairness.

## Chapter 4

# SUBCONTRACTOR BIDS

*The Ron Engineering case revolutionized the law of bidding between general contractors and owners. A few months after that decision, the following case extended and adapted the new principles to bidding between general contractors and subcontractors.*

**PEDDLESDEN LTD. v. LIDDELL CONSTRUCTION LTD.**
Supreme Court of British Columbia; September 1981

In 1979, Peddlesden Ltd., a masonry contractor, submitted through the local bid depository a bid for the masonry work on a school. Liddell Construction Ltd. used Peddlesden's bid in submitting its own general bid to the owner and nominated Peddlesden as its subcontractor. Liddell was awarded the contract and advised Peddlesden by letter of intent that it was awarded the subcontract for all the masonry work.

Together with its bid, Peddlesden had filed a bid bond made in favour of the prime contractor as required by the bid depository. However, the subcontractor failed to seal the bond. When Liddell discovered this omission, it wrote to Peddlesden and stated that since the bond was incomplete, Liddell would not use Peddlesden's bid.

Liddell had the masonry work done on a cost-plus basis by another firm, one that had not submitted a bid for the job. Peddlesden sued, claiming damages for breach of contract. In its defence, Liddell pleaded that:

- the bid bond was invalid, as it was not properly executed under seal as required, so the subcontractor's bid was also invalid;
- the subcontractor's bid expired two days before Liddell purported to accept it in its letter of intent; and
- if Peddlesden's bid had been accepted in time, it would still be necessary to sign a construction contract and, since there was no such contract, there were no contractual ties between the parties when Liddell cancelled the subcontract with Peddlesden.

Justice Ruttan examined the nature and legal effect of the bid submitted by Peddlesden. He relied on the decision of the Supreme Court of Canada in the *Ron Engineering* case.

Liddell, when it argued in court that the bid had expired, relied on the fact that it did not notify Peddlesden within the 30-day period that its bid had been accepted. Counsel quoted a general principle of common law from *Goldsmith on Canadian Building Contracts*:

> In the absence of any express provision to the contrary, an acceptance is not effective until it has been communicated to the offeror and may be withdrawn before that time.

Bidding contracts are different, countered the judge. In *Ron Engineering*, Justice Estey of the Supreme Court distinguished the bidding contract or Contract A from the construction contract itself, or Contract B.

The bidding contract arises when the contractor or subcontractor submits its bid. "The significance of the bid in law is that it at once becomes irrevocable if filed in conformity with the terms and conditions under which the call for tenders was made and if such terms so provide", said Justice Estey.

The *Ron Engineering* case had to do with a general bid directed to the owner. Justice Ruttan now extended the *Ron Engineering* principles to subcontractor bids submitted to the general contractor. When the subcontractor submits its bid, a bidding contract (Contract A) is automatically created. Two further requirements must be satisfied to make it binding on the general contractor:

- the owner must accept the general contractor's bid; and
- that acceptance must be made within the period during which the bid is irrevocable.

Whether the subcontractor is notified within that period or not makes no difference. It is the *owner's* acceptance of the general contractor's bid which makes the subcontract binding, not the notification of acceptance.

Counsel for the general contractor also raised the issue that there was no contract between the parties, since there remained many details to work out, and the construction contract (Contract B) was required to spell those out. The court, however, found no evidence that the legal relationship between the parties was conditional on further negotiations and was a more formal construction contract.

There still remained the question of Peddlesden's failure to affix its seal to the bid bond. When Liddell notified Peddlesden of the omission of the seal, Peddlesden immediately offered to affix the seal to the bond. Its offer was refused. Liddell had had previous experience with Peddlesden, and on three or four occasions Peddlesden had failed to execute the bid bonds. Liddell, therefore, did not want to have anything further to do with this subcontractor.

Certainly, the failure to produce a bid bond renders the bid invalid, decided the court, since the bond is a required part of the procedure in the bid depository system. However, though Liddell may have had its suspicions, there was no evidence that Peddlesden was trying to avoid responsibility under the bid bond by not putting the company seal on it. The court was satisfied that this was a mere slip in executing the document. It could have been corrected immediately and would not have affected the rights and obligations of the parties.

Justice Ruttan found that Liddell had breached the Contract A and that Peddlesden was entitled to damages. The subcontract was tendered and originally accepted by the general contractor for the sum of $53,864. According to Peddlesden's estimate sheet, labour and materials cost $41,678, and the difference of $12,136 represented overhead and profit. The court made some adjustments to that figure and awarded Peddlesden $11,058 for overhead and lost profit, with costs.

---

*Thus, the principles first announced in Ron Engineering were broadened and changed to include subcontractors.*

*The general contractor invites subcontractors to submit offers for parts of the project. The reason for the invitation is that the general contractor wishes to develop a list of subcontractors which it will "carry" in its prime bid to the owner. The mere submission of an offer by subcontractors does not give rise to Contract A. Subcontractor offers — even for the same discipline — will differ. Besides, neither party intends that the invitation and the corresponding offers give rise to contractual obligations.*

*Contract A for contractors/subcontractors can only arise in circumstances where the contractor itself is responding to an owner's offer to enter Contract A by submitting a compliant bid. Then, if the general contractor carries a subcontractor for a particular part or division of the work, the act of carrying the subcontractor creates the bidding contract — Contract A. The subcontractor knows that the general contractor will rely on its offer and that the general contractor's bid to the owner will be irrevocable. The subcontractor's carried offer is also irrevocable and that promise can be enforced through the contractor/subcontractor version of Contract A.*

*Normally, at common law, acceptance of an offer does not create a binding contract unless the person who made the offer is informed that his or her offer has been accepted. The court in Peddlesden decided that the bidding process was different: the contract between the general contractor and the subcontractor arises automatically as soon as the owner accepts the general contractor's bid.*

*Whether the general contractor actually tells the subcontractor "I accept your bid" makes no difference.*

---

As the Contract A/Contract B concept became firmly established in the construction industry, subcontractors started to test its implications, as the following case shows.

## WESTGATE MECHANICAL CONTRACTORS LTD. v. PCL CONSTRUCTION LTD.

Supreme Court of British Columbia; May 1987

In 1982, several general contractors, including PCL Construction Ltd., were invited to submit bids for the construction of a building in Vancouver. PCL in turn invited Westgate Mechanical Contractors Ltd. to bid on the mechanical portion of the project. Westgate telephoned in its bid of approximately $3.4 million.

PCL used Westgate's price in compiling its own bid to the owner, nominated Westgate as its proposed mechanical subcontractor and duly submitted the bid.

None of the bids were accepted by the owners, so the project architect negotiated with the three low bidders for a lower price. PCL agreed to reduce its price by $285,000 and made some other concessions. This was acceptable to the owners, and a memorandum of agreement was executed.

PCL advised Westgate of this and asked the subcontractor to reduce its price by approximately $120,000. Westgate refused. It took the position that because it was nominated by PCL as mechanical subcontractor in the original bid, PCL was bound to employ it at the original price. PCL disagreed and hired another subcontractor. Westgate sued.

In court, Westgate argued that as soon as PCL signed the agreement with the owner, a subcontract came into existence between Westgate and PCL, by law and by the practice in the industry. PCL was, therefore, obliged to employ Westgate.

PCL, on the other hand, maintained that the agreement it signed with the owners had nothing to do with its bid but was a separately negotiated contract. The subcontractor countered that the memorandum of agreement did no more than amend PCL's bid and that the owner had simply accepted the bid in its amended form.

The court agreed with Westgate that, *in principle*, when the owner accepts the general contractor's bid, a contract is created between the contractor and the subcontractor nominated in the bid, and for the bid price. However, the court could not conclude from the evidence that PCL's bid was, in fact, accepted by the owner:

> I find as a fact on the evidence that the owner accepted none of the tenders and what, in fact, happened was that a contract for the construction of the building was negotiated with the defendant on different terms than tendered, including a substantial reduction in price, shortening of completion time, and losing a bonus. These matters are of substance and not a minor adjustment of prices within a tender.

Westgate's claim was dismissed with costs. The British Columbia Court of Appeal affirmed the decision.

---

*The Westgate decision could provide irresistible temptation to owners and, perhaps, to contractors.*

*Consider the situation where an owner issues bid documents, including the usual "lowest or any tender not necessarily accepted" clause, obtains bids and then allows all the bids to expire or rejects them all. Having obtained valuable pricing information, the owner may now negotiate a more favourable deal with one of the bidders.*

*Westgate appears to free the owner from obligations to its former bidders, as well as the contractor from obligations to its subtrades.*

*But appearances are not everything. It could be argued that the owner should have negotiated with the low bidder, not the low three. Setting that aside, if the owner had not come up with a satisfactory negotiated price, all of the bids would have been rejected. When it did come up with a price, the scope of work had changed and, in effect, the offer that was being accepted by the owner was not the bid. When the contractor came to Westgate, it was a parallel negotiation to the one which had occurred between the owner and PCL. If PCL could not negotiate a satisfactory price after a change in work scope with Westgate, then it should have been entitled to reject the subtrade bid and move on, which it did.*

*The next case also raises a question of professional ethics. A general contractor takes advantage of a subcontractor's special knowledge and expertise to prepare its bid to the owner but then does the work with its own forces, ignoring the subcontractor. The contractor may be unfair and unprofessional — but are its actions illegal?*

*The case is also of interest because in it the court was asked to extend that Ron Engineering doctrine further, to include the case where the subcontractor was **not** nominated in the general contractor's bid to the owner, but only used its **price** in making up the bid.*

## RON BROWN LTD. v. JOHANSON

Supreme Court of British Columbia; August 1990

Dwain Johanson, a general contractor, and Ron Brown, a mechanical subcontractor, both wanted to work on the rebuilding of a sewage treatment plant in Vernon, B.C.

Brown gave his subcontract price to all general contractors bidding on the job. Johanson turned out to be the lowest bidder. Brown had the lowest price for the mechanical work, but Johanson decided to do the work with his own forces.

Brown then sued Johanson for the loss of profit he would have made had Johanson hired him. He accused the general contractor of using the information and price contained in Brown's bid without declaring that it was Brown's, and without disclosing that he intended to do the mechanical work himself.

"The question of fact whether Johanson took advantage of Brown's skill and research is not difficult to resolve; he did", said Justice Hamilton after a review of the evidence.

Brown telephoned his bid information to Johanson the day the bids closed. Johanson did not designate Brown as a subcontractor, nor did he specify the use of his own forces for the mechanical work. At the time of bidding, he was required only to bid a lump sum and not to give a breakdown. When he was found to be the low bidder, he gave his breakdown, leaving the space for the nomination of a mechanical subcontractor blank.

The court found that any research he did on his price for the mechanical work was done after he learned that his lump-sum bid was the lowest. One of the factors that persuaded Justice Hamilton was the conspicuous absence from Johanson's evidence of any telephone records indicating long-distance calls to suppliers.

The question of law then facing the court was whether a general contractor is bound to give subcontract work to the lowest compliant bidder if the contractor has not declared that he may use his own forces. Said the judge:

> Bid-shopping, in its various forms, is frowned upon in the construction industry, but is it against the law? A contractor who indulges in such practices may be subject to commercial sanctions referred to in the trade as 'black-balling,' but does he have to pay damages?

Had Johanson nominated Brown in his bid, he could have been forced to hire him — following the decision in *Ron Engineering,* a subcontract would have been created automatically once the owner accepted the general contractor's bid.

Brown asked the court to extend that doctrine even further, to include the case where the subcontractor was *not* nominated in the bid documents, but the subcontractor's *price* had been used in making up the bid.

The court refused to extend the law in this way. Could Johanson have forced Brown to do the work, not having named him as a subcontractor? Or, in general terms, is a subcontractor bound by an implied contract to do the work if not nominated by the general contractor who has the lowest bid?

The court found that the answer to both questions is negative; therefore, if a general contractor could not force the subcontractor to do the work, then the subcontractor could not force the general contractor to hire it.

Further, decided the judge, there is no *implied contract* between a general contractor and all those that submit it subcontract prices that it must nominate the lowest compliant bidder. The contractor, if it wants to, may make such an agreement with a subcontractor, but such an agreement cannot be implied by the courts.

> **Implied Contract**. A contract not created by express words but inferred by the courts either from the conduct of the parties or from some special relationship existing between them.

Brown obtained the names of all the contractors that had taken out plans on the Vernon project, and he phoned his price to each of them. Would a general contractor that had not solicited a price from Brown be obliged to nominate Brown if its bid turned out to be the lowest? Obviously not. If submitting an unsolicited price is an offer, then acceptance can be implied only when the general contractor nominates the subcontractor.

Since Brown failed to prove any express or implied agreement with Johanson, the court dismissed his claim in contract. However, Brown had also based his claim, in the alternative, on an allegation of *fraud* and *unjust enrichment*.

To prove fraud, Brown was required to prove that Johanson had told him that he did not intend to do the mechanical work with his own forces, and that the statement was made fraudulently. Brown could not prove this. If Johanson misled Brown, it was at a time when Johanson himself was uncertain as to what he would do because he was bidding on three jobs at the same time.

To succeed in his claim of unjust enrichment, Brown had to establish that he was deprived of something that enriched Johanson, who should have been compelled to turn over his gain to Brown because he had no legal right to keep it. The court found that Johanson did not deprive Brown of anything. Brown had intended from the start to bid the job, and he did his pricing not just for Johanson. No doubt, he expected to get the job if his price was low, but he had no *legal* right to expect that.

Although Johanson deprived Brown by failing to nominate him and enriched himself by doing the work with his own forces, there was no contract to nominate Brown, and so there was no legal reason why the contractor should turn over his profit on the mechanical work to Brown. The claim based on unjust enrichment, therefore, failed.

However, Johanson did not leave the court entirely unscathed. The trial took 13 days. The judge estimated that ten of those days were taken up in trying to decide whether Johanson had used Brown's price, and on this one issue Brown won. The court, therefore, decided that Brown should recover from Johanson one-third of his taxable legal costs.

---

*At this stage in the development of Contract A between contractors and subcontractors, bid shopping was seen as merely anti-social. As a result of the Naylor decision, bid shopping has taken on an entirely different flavour. If Ron Brown were to make its case today, the result might be different.*

*The Alberta Court of Queen's Bench embarked on a different road than the court in Ron Brown.*

## BATE EQUIPMENT LTD. v. ELLIS-DON LIMITED

Court of Queen's Bench of Alberta; July 1992

In 1982, the Edmonton School District No. 7 called for bids for a school building. Subcontractor bids, submitted through the Edmonton Bid Depository, closed on March 23rd, and bids from general contractors, two days later.

The general contract was awarded to Ellis-Don Limited, while the elevator subcontract went to Dover Corporation (Canada) Ltd. Soon after the closing of the subcontract bids, Douglas Bate, the principal of another elevator company, Armor, found out that his company was the second lowest bidder but that Dover, the lowest bidder, had qualified its bid.

The offending words in the Dover bid were as follows:

> We are pleased to submit our quotation to supply and install two (2) variable voltage geared passenger elevators and one (1) oildraulic passenger elevator, generally in accordance with the plans, specifications addenda, utilizing our manufacturing standards.

Bate lodged a formal complaint. The management committee of the depository upheld it. The committee decided that Dover's bid was informal, and the committee instructed the depository staff to disqualify it. Ellis-Don then showed Armor as its proposed elevator subcontractor and used Armor's price in its own bid.

Dover, however, appealed immediately to the architect to have its bid reinstated. It complained that its bid form was a standard form it had used for several years, and it argued that its bid was not qualified.

The architect was familiar with Dover, which had earlier helped to draft generic elevator specifications for the project. It did not think that there was a defect in Dover's bid, nor did the qualification give Dover an unfair advantage over other bidders. Therefore, the architect recommended that the owner request that Ellis-Don show Dover as the elevator subcontractor.

The owner accepted the revised bid by Ellis-Don showing Dover as subcontractor. A formal construction contract was signed. Bate sued for breach of Contract A.

At the trial, Justice Dea said:

> The architects' motives were proper. Dover was a competent subcontractor and the architects were aware of that. The best interest of the owner was to get a competent subcontractor at the lowest price. That is what Dover was and that was what Armor was not.

However, the main issue at trial was whether Armor entered into Contract A with Ellis-Don and, if so, what were the terms of this contract? Justice Dea found that, beyond question, Contract A was formed between Armor and the general contractor when Armor submitted its bid. The terms of that contract were far less certain.

Armor's principal argument was that Ellis-Don accepted its bid for the elevator subcontract when it carried Armor's name and price in its bid for the general contract to the owner. In *Peddlesden v. Liddell Construction Ltd.*, the court decided that once the owner accepted the general contractor's bid, the contract between the general contractor and the subcontractor became binding. The contract did not depend on whether the general contractor communicated to the subcontractor that its bid had been accepted:

> ... while communication of acceptance is normally required to make a binding contract between the parties, in tender contracts [it is the acceptance of the bid by the owner] not the subsequent communication of acceptance that creates the binding contract.

Justice Dea did not think that the *Ron Engineering* case went so far. The normal law of contract obligates a party accepting a contract to give the other side notice of acceptance. Further, a number of provisions in the bid documents worked against such a view of acceptance. The Instructions to Bidders, for example, stated:

> 7.1 The owner reserves the right to accept the tender which is deemed most advantageous. The lowest or any tender will not necessarily be accepted.
> 7.2 After acceptance by the owner, the architect on behalf of the owner will issue to the successful bidder a written notice of acceptance.

These contract provisions applied as between the owner and general contractor, but the general conditions of the contract also made them apply as between the general contractor and the subcontractor.

Having considered various precedents, Justice Dea concluded that there was no compelling authority that the mere fact of carrying a subcontractor's name and bid constituted acceptance. On the contrary, he found abundant reasons to reject such a rule.

In particular, the contract documents called for Ellis-Don to disclose its "proposed subcontractors", and the general contractor knew that the owner was entitled to reject the proposed subs. In view of this, no general contractor would want to be under an obligation to enter into a Contract B with a sub before its own bid had been accepted and the subcontractors approved by the owner.

Accordingly, concluded the judge, it follows that, while Armor and Ellis-Don formed a Contract A, the contractor did not accept Armor's bid either expressly or by virtue of carrying Armor's bid. As a result, at no time did Ellis-Don become obligated under Contract A to enter into Contract B with Armor.

The court also considered the effect of the bid depository on the dispute. Did these rules and the decision of its management committee which disqualified Dover apply to Armor's contract rights under Contract A? And, if so, were those rights breached when Ellis-Don entered into Contract B with Dover rather than with Armor?

The Instructions to Bidders provided that

> ... the services of the Edmonton Bid Depository will be required on this tender for prime subtrades and sub-subtrades in accordance with the rules of the depository.

Therefore, the bid depository rules applied to Armor and all other bidders as part of the Contract A between Ellis-Don and the subtrades. On the face of it, Ellis-Don breached Contract A when it entered into a contract with Dover because Dover had been disqualified under the depository rules.

However, the court found a conflict between this provision of the bidding instructions and other contract documents. For example, GC 10.3 stated:

> The Owner may, for reasonable cause, object to the use of a proposed Subcontractor and require the Contractor to employ one of the other subcontract bidders.

Further, as mentioned before, the contract gave the owner the right to reject any or all bids.

The contract clearly specified which documents have priority over others. Thus, when the bid depository rules and rulings conflicted with the other provisions of the contract, those *other* provisions prevailed and took priority. They authorized the owner to reject the rulings of the depository management committee, as it in fact did when it accepted Dover.

Therefore, concluded Justice Dea, the actions of Ellis-Don in following the owner's order to use Dover, and accepting Dover's bid and entering into Contract B with it, did not breach Contract A between Ellis-Don and Armor.

Finally, even if Ellis-Don had breached Contract A, the damages would be limited. Such a breach of Contract A would only mean that Ellis-Don ought not to have signed a contract with a disqualified bidder. The general contractor was never under an obligation to accept Armor's bid or the bid of any of the other elevator bidders.

In short, the breach would be, at best, technical breach — perhaps entitling Armor to the costs it incurred in preparing its bid.

---

*In the following case, the Ontario Divisional Court wrestled with the question of communication of acceptance and came down on the side of traditional contract law.*

*Also, a deficient offer (call for bids) or acceptance (bid) cannot be the basis of a bidding contract.*

## SCOTT STEEL (OTTAWA) LTD. v. R.J. NICOL CONSTRUCTION (1975) LTD.

Ontario Divisional Court; February 1993

This case revolved around a single issue: was there, or was there not, a binding bidding contract (Contract A) between the general contractor and the subcontractor?

In 1985, the Carleton Board of Education invited bids for the construction of a new school. The contract documents contained the following provisions:

> 10.2 The Contractor agrees to employ those Subcontractors proposed by him in writing and accepted by the Owner at the signing of the Contract.
> 10.3 The Owner may, for reasonable cause, object to the use of a proposed Subcontractor and require the Contractor to employ one of the other subcontract bidders.
> 10.5 The Contractor shall not be required to employ as a Subcontractor a person or firm to whom he may reasonably object.

The Instructions to Tenderers specified that "the lowest or any tender shall not necessarily be accepted", and also made the following provisions:

> 12.1 Within 24 hours of tender closing, the tenderer shall submit to the Architect, Section 00430 List of Subcontractors ...

12.2 The Contractor will not be allowed to substitute other Subcontractors in place of those named in his tender without written approval from the architect.
12.4 The Owner reserves the right of refusal on any Subcontractor.

R.J. Nicol Construction (1975) Ltd., a general contractor, requested Scott Steel (Ottawa) Ltd. — among others — to provide a quote for supplying structural steel on the project. Scott quoted its price by telephone.

Nicol used Scott's quotation in its bid to the owner and listed Scott as its subcontractor. Nicol was awarded the general contract but decided to award the subcontract to another firm. It believed that Scott could not perform the work according to the project schedule, which was of paramount importance.

The trial judge found no evidence of bid shopping — the contractor paid more to the replacement subcontractor — nor did he find an improper motive for the change. He, therefore, found that there was no contract between the subcontractor and the general contractor. The reasons follow:

- The subcontractor's bid was not complete as it did not properly set out the scheduling.
- Articles 12.2 and 12.4 quoted above proved that the mere submission of a bid did not create a contract between the general contractor and the sub when the owner accepted the general contractor's bid.
- This view was further fortified by sections 10.2, 10.3 and 10.5 also quoted above.
- The subcontractor admitted that it knew the project could not be completed on time if the structural steel work was completed in accordance with its schedule.

In *Ron Engineering*, Justice Estey made it clear that there might be situations where a contract would *not* arise when the contractor submits its bid. A bid may be so deficient as not to amount to a bid at all. In such cases the owner could not "snap up" such a bid; thus, there would be no bidding contract.

Similarly, there may be situations where the call for bids is so deficient that it does not amount to an offer that a bidder can accept and so create a bidding contract, said Madam Justice Charron of the Ontario Divisional Court while considering Scott's appeal.

Nicol's call for bids to subcontractors contained no particular terms. Scott, in its telephone bid, did not stipulate any terms other than the price. It is difficult to see, said the judge, how this process — in isolation — could give rise to a binding contract.

The subcontractor argued, however, that all relevant terms and conditions were set out in the owner's call for bids and that the court should imply them as part of Scott's bid since both Scott and Nicol knew the owner's specifications. Scott relied mainly on the decision of the Supreme Court of British Columbia in *Peddlesden*.

Justice Charron agreed that the *Ron Engineering* principles *should* be applied to the bidding process as between the trades and the general contractor. However, she could see no reason why the normal rules relating to acceptance *should not* apply.

It is important, continued the judge, to maintain the sanctity of the bidding process. The rules must enable the general contractor to rely on subcontractor quotes, since the general contractor in turn binds itself to the performance of those terms in its bid to the owner. This process would not suffer if the contractor was bound in its contractual relationship with the subtrade only if it (a) had the contract with the owner and also (b) let the subtrade know that its bid was accepted.

*Ron Engineering* clearly stated that the bidding contract is a *unilateral contract*. Once the subcontractor submits a bid, the bid becomes irrevocable for a certain period of time. If the general contractor accepts its bid, the sub may be obligated to enter into a construction contract — Contract B.

But it is in the nature of a unilateral contract that the person who receives the bid, in this case Nicol, is not obliged to accept the bid — a position that seems to contradict *Ron Engineering*. Nor is he or she obliged to enter into a Contract B unless he or she accepts the bid *and* lets the bidder know that he or she has accepted it, as the usual rules of contract law require him or her to do.

Justice Charron quoted with approval Justice Dea of the Alberta Court of Queen's Bench in *Bate Equipment*:

> In my view, there is no compelling authority that the mere fact of carrying the subcontractor's name and bid constitutes an acceptance, conditional or otherwise, of the bid.

In summary, the court came to the following conclusions:

- The principles of *Ron Engineering* applied to the bidding process between Scott and Nicol.
- The specifications and requirements set out by the owner and known to both Scott and Nicol were implied in Scott's bid.
- Scott knew that Nicol could well rely on its bid in the preparation of its own bid to the owner.
- Scott knew that if the owner accepted Nicol's bid, Nicol would have to perform that portion of the contract in accordance with the terms of Scott's bid.

- It is reasonable to conclude that the parties intended Scott's bid to the owner to remain irrevocable for the same period as Nicol's.

At that point, the bidding contract — Contract A — was formed.

However, the contract was unilateral in nature and did not impose any obligations on Nicol to accept Scott's bid or to enter into Contract B:

- Nicol would only be liable to enter into Contract B if and when it accepted Scott's bid and communicated this acceptance to Scott.
- Putting Scott on its list of *proposed* subcontractors did not constitute acceptance of its bid.
- The owner's acceptance of Nicol's bid made no impact on the relations between Scott and Nicol — the owner's actions affected only the rights between Nicol and the owner, and Nicol was still free to change subtrades with the owner's consent.

Nicol never accepted Scott's bid so as to create contractual obligations between the two of them, concluded Justice Charron. Justice White agreed, while Madam Justice Greer dissented. Thus, by a majority decision, the court dismissed Scott's appeal.

---

*In this case, the critical facts are:*

- *that the subcontractor provided a bid by telephone just before prime bids closed;*
- *that the contractor named the sub and carried its price before it had an opportunity to confirm that the sub's bid was in accord with the plans and specifications; and*
- *that when the contractor attempted to confirm the telephone bid, it discovered that the sub's bid was not "plans and specs" in particular with regard to scheduling.*

*The court held that the listing of the sub in the contractor's bid and the owner's acceptance of the bid were not sufficient to bind the contractor to the sub. A binding subcontract would exist only if the contractor **formally communicated** its acceptance of the subtrade bid within the period of bid irrevocability.*

*Before Scott Steel, the leading subtrade case on the point, namely Peddlesden, recognized that naming a sub was a serious commitment. In principle, it meant that the sub will get the subcontract unless the contractor is unsuccessful or unless the owner exercises its rights to reject particular subs.*

Naming the sub in the prime bid represented a conditional acceptance of the sub's offer; this acceptance becomes unconditional when the owner accepts the prime bid. There is no need for further communication between contractor and sub. At least, that is what Peddlesden established.

When sub pricing is taken by telephone, the process is fluid. Contractors are preoccupied with pricing and with the fact that they are running out of time. Rarely do they get detailed offers from subtrades, and even more rarely do they have an opportunity to assess those offers.

Subs are preoccupied with getting carried and with submitting their prices as close to the closing deadline as possible to avoid their prices being shopped. When contractors receive lead letters, after the prime bid has closed, the letters often contain surprises.

If the lead letter from the subcontractor shows that there was no "meeting of minds" between the sub and the contractor, the conditional subcontract is over, and the contractor is free to get another trade. Not only is that fair, it is what subs and contractors would normally expect.

In this sense, the decision of the court in Nicol was the right one. But, by leaving open-ended the question of whether a contractor had any obligation to the trades it "carried" it was problematic. Was a contractor now free to position acceptance of the trade's offer against a price concession?

## VIPOND AUTOMATIC SPRINKLER CO. LTD. v. E.S. FOX LTD.

Ontario Court (General Division); April 1996

In 1992, Vipond Automatic Sprinkler Co. Ltd. submitted a bid to the mechanical subcontractor, E.S. Fox Ltd., for the sprinkler installation on a project in Cornwall, Ontario. Fox submitted its own bid to the general contractor. However, the mechanical bids were grossly over budget, so the owner rejected all of them.

After several additions and deletions, Fox filed a revised bid. The bid included Vipond's price of $139,900, which was the lowest bid for the sprinkler work on the original bid.

Fox's revised bid was successful. The construction contract included an addendum requiring a change in the sprinkler system using a cheaper type of pipe. Fox, therefore, sent a fax to the three lowest sprinkler bidders, including Vipond, asking for a revised price.

Desco Fire Protection, which had submitted to Fox the second lowest bid in the initial bidding process, now offered a revised price of $127,800 for using the cheaper piping. Vipond informed Fox that there was no need for it to revise its original bid since the piping it was proposing to use was already in accordance with the addendum. Fox then awarded the sprinkler contract to Desco.

After it found out about the award, Vipond offered Fox cost savings of $20,000. When this failed to change Fox's mind, Vipond sued for loss of profit. It claimed that Fox had breached a verbal agreement to award the sprinkler contract to Vipond.

Vipond based its claim of breach on two telephone conversations it had with a representative of Fox after the initial bid. The representative, it alleged, had advised Vipond that it would be issued a purchase order for the sprinkler work. Fox denied that any agreement with Vipond had been reached.

The court concluded for several reasons that the version of what happened as presented by Fox was closer to the actual events:

- It would not have made commercial sense for Fox to make a commitment to Vipond before sending out the fax asking for revised prices.
- In subsequent communications between Vipond and Fox, there was no mention of the alleged prior agreement.
- Vipond's evidence was not unequivocal: it indicated that further discussions would have to take place.

There were other reasons that led the court to conclude that Vipond had the *impression* that a purchase order would be forthcoming, but not an *agreement*, either express or implied. Fox never actually promised Vipond the contract, so there was no breach. Vipond simply favoured a particular interpretation of the telephone calls, decided Justice Chapnik.

As an alternative to its claim of breach of contract, Vipond claimed that Fox treated it unfairly during the bidding process and thereby breached the duty of fairness it owed to the bidders.

The bid documents contained the common privilege clause stating that "the lowest or any tender will not necessarily be accepted" and gave the owner the right to reject any and all bids.

Justice Chapnik found that, based on precedent cases such as *Scott Steel (Ottawa) Ltd. v. R.J. Nicol Construction (1975) Ltd.*, the fact that the owner accepts a contractor's bid does not mean that the subcontractors and sub-subcontractors are also accepted. Therefore, the subcontractor is free to change its subtrades.

"[The Scott Steel decision], in my view, reflects the prevailing law in Ontario and is in conformity with commercial reality", said Justice Chapnik and summarized her findings:

> It is my finding that the mere carrying of [Vipond's] price in the revised bid did not constitute an acceptance, conditional or otherwise, of [Vipond's] bid nor did it create an obligation on the part of [Fox] to present Desco's revisions to Vipond.

There was, furthermore, a special aspect to the conditions surrounding the bidding process in the case of Vipond and Fox. Justice Chapnik put it this way:

> In the particular situation here, an intervening circumstance altered the nature of the subcontract. By all accounts, after the initial tender, the bidding procedure degenerated into a "free for all" without discipline or any formalized re-tendering process.

The owners were pushing for the project to start, while the construction industry as a whole was slow, the margins close, and everyone was "constantly sharpening their pencils to go after a job".

The judge relied on a decision reached under similar circumstances by the British Columbia Court of Appeal in *Westgate Mechanical Contractors Ltd. v. PCL Construction Ltd.*

In that case, Westgate submitted a bid to PCL which named Westgate as its proposed mechanical subcontractor. The owner initially rejected all of the bids. The court found that when the owner finally awarded the contract to PCL, it was a new contract, negotiated on terms different from those underlying the original bidding process. Therefore, Westgate's claim that it had a bidding contract with PCL was invalid.

Similarly, in the case of Vipond, the informal circumstances in the re-tendering process rendered the subcontract between the general contractor and Fox a negotiated agreement rather than a continuation of the original bid.

There was no evidence that Fox had acted improperly, in bad faith or with unfairness. There was no violation of the bidding process, no bid shopping — therefore, no wrong-doing whatsoever:

> Sending the same fax to all three low initial bidders, proceeding at arm's length, and in all of its actions, [Fox] acted in good faith and in a fair manner towards [Vipond].

In short, the owner's or general contractor's acceptance of Fox's bid did not impose a legal obligation on Fox to accept Vipond's bid:

> Nothing in the contract documents required the successful contractor to enter into an agreement with any proposed subcontractor; and [Fox] as mechanical contractor was entitled to accept any bid it deemed most favourable to its interests.

The court dismissed Vipond's claim and awarded Fox costs amounting to $15,000.

---

*So far, the cases have dealt with a situation where the subcontractor, contrary to its expectations, is rejected by the general contractor. The following case deals with the other side of the coin: the general contractor going after a sub reluctant to sign a construction subcontract.*

## STUART OLSON CONSTRUCTORS v. NAP BUILDING PRODUCTS

Supreme Court of British Columbia; February 1997

Stuart Olson Constructors is a Vancouver general contractor. In early 1994, it was bidding on the construction of a high rise apartment complex. NAP submitted the low subcontract bid of about $680,000 for the supply of windows for the building. Olson carried NAP as its window supplier.

In early May, NAP requested a letter of intent from Olson confirming that it was indeed chosen as the window supplier. On May 4th, Olson gave NAP a letter of intent conditional only on the owner entering into a contract with Olson.

Then, on May 16, the general contract was awarded to Olson.

As often happens, the call for trade prices did not include the form of subcontract. But Olson and NAP had done business together before. Accordingly, Olson prepared a draft subcontract based on an earlier document and sent it to NAP for its review.

Windows coming later in the project, there was no real urgency to finalize the form of subcontract. NAP made some amendments to the draft, and then one of its officials signed and returned it to Olson in October 1994.

In mid-November 1994, the parties met and discussed the subcontract. The only issue of any consequence was a labour clause. Following the meeting, NAP took some steps to order materials but changed its mind and repudiated its dealings with Olson on November 25.

In the face of the repudiation, Olson went back to the market and obtained three bids for the window work. The lowest of those three bids was approximately $169,000 higher than the price that had been received from NAP. Olson awarded the work to the new window trade and then sued NAP.

At trial, *Ron Engineering* was invoked to examine the relationship between the parties. NAP argued that there was no relationship between the parties and no Contract A. As a back-up position, it claimed that there never was a Contract B (subcontract for the windows) and that the best Olson could do was recover damages for breach of Contract A, which would be minimal.

The trial judge would have none of NAP's position. He concluded that Contract A was properly formed when the subtrade bid arrived from NAP. Under Contract A, NAP was obliged to sign Contract B and perform the work if it was awarded. On the facts, Contract B had been awarded and was repudiated by NAP.

Olson claimed the spread between NAP's price and the price it later had to pay for the windows ($169,000) plus the cost of re-tendering ($7,500).

Realizing it had lost the Contract B argument, NAP argued that the measure of damages was the spread between its price and the next low bidder at the time of the original bid in early 1994. NAP also argued that it had recommended some changes in the work that would have reduced the contract price by $13,500. It also claimed that Olson had upgraded the quality of the window installation to the tune of $21,000. NAP wanted deductions for these two items.

The trial judge held that Olson's damages should be calculated based on the re-tender, not on the original subtrade pricing, which was not available when NAP repudiated. The trial judge reduced Olson's damages by $13,500, finding that the NAP proposed changes were valid. The trial judge rejected NAP's claim that Olson had upgraded the windows, finding that this had not occurred.

Olson's claim for the cost of re-tendering was not allowed. The trial judge found that the claim had not been properly proven.

Although it may appear that recovering the cost of re-bidding is a double recovery, an examination of the circumstances would suggest it is not. Olson conducted a bid process when the overall job was bid and received pricing, including that from NAP. By the time NAP repudiated, one infers that all of the former subcontract bids would have expired, which meant Olson had to go back to the market. Olsen anticipated running the bid process once. Although it could not prove a loss in court, it is clear it had to run a second bid process as a direct result of the default of NAP.

In the final analysis, Olson was awarded approximately $150,000 plus interest and costs.

---

*There is a two-pronged moral to this story.*

*The first prong is that it is tough to argue with the umpire after a swinging third strike.*

*Olson had relied on the NAP price (see Gloge in Chapter 6). It notified NAP that it had been awarded a subcontract (Scott Steel). If strike three was needed, it lay in the action of a NAP representative, who had signed the draft subcontract and returned it to Olsen.*

*The second branch of the moral to this story is that damages have to be proven on the balance of probabilities (more likely to have been suffered than not suffered) in order to recover them. Often, a party concentrates so much on liability that it hasn't done its homework by the time the court says "yes". A hard fought win can be unrewarding if one has overlooked strong proof of damages.*

## NAYLOR GROUP INC. v. ELLIS-DON CONSTRUCTION LTD.

Supreme Court of Canada; September 2001

In 1991, the Oakville-Trafalgar Memorial Hospital called for bids for the construction of an addition and renovation of a hospital. Ellis-Don Construction Ltd., a general contractor, invited Naylor Group Inc. to bid for the electrical work on the project.

At the time, Ellis-Don had a continuing dispute over bargaining rights with the International Brotherhood of Electrical Workers (IBEW). After many years of wrangling, the dispute came before the Ontario Labour Relations Board (OLRB) in 1990. The ruling was still reserved at the time of bidding.

Naylor had an in-house union, not affiliated with IBEW. Ellis-Don assured Naylor that this would not be a problem.

Naylor submitted a bid, $5,539,000, through the Toronto Bid Depository. The bid was the lowest, and Ellis-Don carried Naylor's price in its own bid for the prime contract. The next lowest bid for the electrical work came from Comstock, an IBEW subcontractor, whose bid was $411,000 higher than Naylor's.

Ellis-Don was the low prime bidder at $38,135,900. If it had carried Comstock's electrical bid instead of Naylor's, it would not have been the low bidder overall, and probably would not have won the prime contract.

The OLRB decision was released on February 28, 1992. It confirmed Ellis-Don's obligation to use only electrical subcontractors affiliated with the IBEW. In the meantime, the Hospital had made various changes to its project. Despite the OLRB ruling, Ellis-Don asked Naylor to submit a price for the contract changes, and again carried Naylor's price in its bid to the owner.

In May 1992, with Naylor still carried in Ellis-Don's list of subtrades, the Hospital awarded the prime contract to Ellis-Don. GC 10.2 of the prime contract (CCDC-2 1982) provided:

> The Contractor agrees to employ those Subcontractors proposed by him in writing and accepted by the Owner at the signing of the Contract.

Ellis-Don indeed offered the subcontract to Naylor but on condition that it align itself with IBEW. Naylor refused. Ellis-Don then offered the work to Guild Electric, an IBEW subcontractor but ineligible under the bid depository rules, for exactly the same amount as had been bid by Naylor. The final subcontract was eventually signed with Guild with a minor price difference.

Naylor sued Ellis-Don for breach of contract and unjust enrichment. It claimed that Ellis-Don had used Naylor's low bid to get the prime contract, then "shopped" its bid to get the work done at a very favourable price. All of this, argued Naylor, undermined the integrity of the bid depository process and breached the terms of Contract A.

At trial, Justice Langdon found that the award of the prime contract to Ellis-Don did not automatically trigger a subcontract between Ellis-Don and Naylor for the electrical work. According to traditional rules of contract formation, there was no contract between the parties until Ellis-Don had *communicated* its acceptance of Naylor's bid to the subcontractor. Ellis-Don had never done that.

In any event, said the judge, if any such contract for the electrical work had come into existence, it was frustrated by the OLRB decision which precluded Ellis-Don from contracting with a non-IBEW electrical subcontractor.

> **Frustration of Contract**. The unforeseen termination of a contract as a result of an event that either renders its performance impossible or illegal or prevents its main purpose from being achieved.

The trial judge allowed Naylor's claim for unjust enrichment, namely the cost of preparing the bid and related overhead, and awarded Naylor $14,560. He also assessed the damages he would have awarded Naylor if he had found breach of contract: the amount of its lost profit on the project, which he assessed at $730,286.

Naylor appealed. In the Ontario Court of Appeal, Justice Weiler proceeded in accordance with the analysis set out in *Ron Engineering*. In exchange for binding itself to an irrevocable bid, she said, Naylor also acquired rights under Contract A. It was not automatically entitled to the award of the subcontract for the electrical work (Contract B), but such an award had to be made unless Ellis-Don or the owner had a reasonable objection, as provided by GC 10.3 and GC 10.5 of the standard form contract.

Justice Weiler held that Ellis-Don's objection to Naylor was not reasonable for two main reasons: (a) Ellis-Don "shopped" Naylor's bid; and (b) Ellis-Don could have persuaded OLRB to allow the contract with Naylor. In the absence of a reasonable objection, Ellis-Don had breached the terms of Contract A with Naylor.

Justice Weiler accepted the trial judge's estimate of Naylor's loss of profit ($730,286). She then discounted this figure by 50% for job site contingencies, and the resulting figure by a further 50% to account for the contingency that the OLRB would not have allowed the contract to be awarded to Naylor, or that Naylor may have had to enter into a sub-subcontract with another electrical subcontractor with IBEW affiliation. Thus, the Court of Appeal awarded Naylor damages for breach of contract in the amount of $182,500.

Both Naylor and Ellis-Don appealed to the Supreme Court of Canada. In September 2001, ten years after the bidding process started, the high court finally settled the dispute.

The Supreme Court, for the first time, made it clear that the Contract A/Contract B approach applies to subcontractor bids just as it does to prime bids. It then addressed five issues:

- Was a Contract A formed between Ellis-Don and Naylor?
- Was Contract A frustrated by the OLRB decision?
- If not, did Ellis-Don breach the terms of Contract A?
- If so, what are the damages?
- In the alternative, is Naylor entitled to recover on the basis of unjust enrichment?

Justice Binnie, writing for the court, also noted: "There lurked in the background to some of [Naylor's] submissions in this Court occasional allegations which seemed grounded in tort, including negligent misrepresentation." However, tort was neither pleaded nor argued in the courts below, so he did not consider it. The message to the industry is "Stay tuned!"

*1. Was a Contract A formed between Ellis-Don and Naylor?*

Naylor contended at trial that Ellis-Don, on winning the prime contract, became *automatically* obligated to it under the terms of Contract A to enter into the electrical subcontract, *i.e.*, Contract B. Justice Binnie found nothing in the bid documents to give rise to such an obligation. On the contrary, the bid documents clearly contemplated the possible substitution of a subcontractor different from the one in the prime bid. GC 10.3 and GC 10.5 of the standard contract were incorporated into Contract A, and were plainly inconsistent with Naylor's theory.

Furthermore, he agreed with Justice Weiler that there could be no Contract B without Ellis-Don communicating to Naylor the acceptance of its subcontract bid.

Justice Binnie pointed out, however, that the prime contractor, under Contract A, had clear contractual obligations to the subtrades it carried in its own bid. The bid documents created Contract A between them. The terms of this contract were drawn from the bid documents which, in this case, included the provisions of the prime contract and the bid depository rules. Among the terms of Contract A was one which required the successful prime contractor to subcontract to the firms whose bids it had carried unless it had a reasonable objection to one of them.

Therefore, when Ellis-Don chose to carry Naylor's bid in its bid to the owner, it committed itself to subcontract the electrical work to Naylor in the absence of a reasonable objection. What is reasonable depended on the facts of the case.

*2. Was Contract A Frustrated by the OLRB Decision?*

Ellis-Don argued that, even if it was bound by Contract A, it was nevertheless relieved of any obligation by the supervening event of the OLRB decision which ruled out Naylor.

Frustration of a contract occurs when a situation arises for which the parties made no provision in the contract, and performance of the contract becomes a thing radically different from that which was undertaken by the contract.

Justice Binnie found that there was no frustration of the Contract A between Ellis-Don and Naylor. The OLRB ruling was a foreseeable outcome. It recognized and affirmed Ellis-Don's obligation to the IBEW, it did not create it. Accordingly, when Ellis-Don approached Naylor to do the Hospital work with non-IBEW workers, and subsequently carried the bid to the owner, it was promising work that it had already bargained away to IBEW.

Ellis-Don was in no better position than someone who sells his house to two different buyers. A court decision upholding the validity of the first contract of sale would not frustrate the second contract. It would simply lay the basis for a claim in damages by the second purchaser. Similarly, the OLRB decision merely affirmed that Ellis-Don had a pre-existing obligation to IBEW, and the loss of the contract entitled Naylor to damages.

Furthermore, the parties to Contract A specifically provided their own test to deal with supervening circumstances which might disqualify Naylor, by means of a flexible exit option based on reasonableness. Thus, the legal issue here was not the doctrine of frustration but the question whether, in light of its conduct under the terms of Contract A, it was reasonable for Ellis-Don to reject Naylor because of its union affiliation.

*3. Did Ellis-Don Breach the Terms of Contract A?*

Justice Binnie had sympathy for the dilemma Ellis-Don was in at the time of bidding. The OLRB decision was pending for about a year. In the meantime, Ellis-Don had either to bid carrying IBEW subcontractors (perhaps unnecessarily) and risk losing major projects, or bid carrying non-IBEW subcontractors (perhaps wrongly) and risk the subsequent wrath of the IBEW and, perhaps, the OLRB.

The problem, said Justice Binnie, was that Ellis-Don tried to solve its dilemma at Naylor's expense. He did not agree with the Court of Appeal's optimism that the OLRB ruling could be abated or bypassed: *"The IBEW, having established the correctness of its position at much effort and expense, could be expected to insist on the fruits of victory."*

Ellis-Don chose to carry Naylor instead of its IBEW affiliated rival, Comstock, and thereby assured itself as low bidder of winning the prime contract.

Ellis-Don's only objection to Naylor was the fact it was not an IBEW subcontractor. In light of Ellis-Don's conduct, however, this was not a reasonable objection: Ellis-Don assured Naylor that its in-house union was no problem, and it carried Naylor's bid in its own bid with full knowledge of the proceedings before the OLRB. It even affirmed its agreement to use Naylor *after* the OLRB ruling. Ellis-Don signed the prime contract, which named Naylor as the electrical subcontractor, two months after it was fully aware of the OLRB decision.

Naylor, not surprisingly, saw Ellis-Don's conduct in a very poor light. It believed that its bid was used by Ellis-Don to obtain the prime contract and used again to secure a substitute electrical subcontract at the same price from Guild, contrary to the rules of the Toronto Bid Depository which formed part of Contract A. Justice Binnie, however, decided to dispose of the case on the narrow contractual ground that Ellis-Don could only extricate itself from Contract A by demonstrating that, in all the circumstances, its objection to Naylor was reasonable. This it had failed to do, therefore it was in breach of Contract A.

*4. What Are the Damages for Breach of Contract A?*

Following well accepted principles, Naylor had to be put in as good a position financially as it would have been in had Ellis-Don performed its obligations under Contract A. The normal measure of damages in the case of a wrongful refusal to contract in the building context is the contract price less the cost of executing or completing the work, that is, the loss of profit.

Naylor's claim for lost profit was $1,769,412. This figure was based on an average mark-up of 12.4% on the contract price, grossed up to an average mark-up of 31.2% on the entire job because of Naylor's demonstrated capacity to squeeze profit out of contract extras.

The trial judge had concluded that this was overly optimistic, and held Naylor to a more realistic mark-up of 11.2% plus other adjustments, producing a figure of $730,286. He noted that Guild Electric had suffered a significant financial loss on the job.

The Court of Appeal had reduced the trial judge's estimate by 50% because it felt the trial judge had failed to take into account a number of relevant features of the unexpectedly adverse conditions on the job site. Justice Binnie agreed with Justice Weiler of the appellate court that the trial judge had failed to take into account the unexpectedly severe site problems, and accepted the reduction of the loss of profit to $365,143. However, the further reduction of Naylor's loss of profit to $182,500 for labour relations contingencies was, in his opinion, too speculative and he increased the damage award to $365,143.

In light of the conclusion that Naylor was entitled to recover damages for breach of contract, there was no need to examine the alternative ground of unjust enrichment.

The *Naylor* case involved the bid depository. So, some contend that *Naylor* will only apply when the subcontractor is sourced through the bid depository.

Although it is always possible that a court will distinguish *Naylor*, the more likely approach — and the one that the industry would be wise to consider — applies the principles of the case to all contractor/subcontractor bid situations - where the prime bid falls under the umbrella of *Ron Engineering*.

What if an owner called bids and stipulated that the mechanical and electrical trades will be sourced through the bid depository and everything else not — a common occurrence. Would a court hold that the carried electrical sub had Contract A while the "carried" drywall sub did not? Our sense is that the terms of Contract A will be a bit different for those two subcontractors — different before the moment of prime bid closing. But, after each is "carried", both subs will be accorded the same comfort:

"The assurance of a subcontract to the carried subcontractor, ... was ...the most important term of Contract A."

So what do we have?

In a bid depository situation, the bid documents are identified and available to the trades. And, the rights and obligations of the parties flow from the bid documents which usually include the bid depository rules and the prime contract. Contract A comes into existence at the moment the bid depository subcontractor is "carried".

In a non-depository process, access by the trades to the prime bid documents is the key to whether Contract A can arise (do the trades have access or not?). If Contract A can arise, *Naylor* identifies its birth as the moment the subcontractor is "carried". After that, Contract A derives its terms from the prime bid documents.

The likely paradigm for Contract A — whether in bid depository or not — is:

- Contract A arises when the trade is "carried';
- without "reasonable" owner or contractor objection, the contractor must award subcontracts to "carried" subcontractors;
- upon award, the sub must perform the work of the subcontract;
- "reasonable objection" must arise after prime bid closing and be significant commercially or contractually;
- no post closing bid shopping.

## SUMMARY

The following general conclusions may be distilled from the decisions summarized in this section:

- Contract A between a subcontractor and contractor arises when the subcontractor submits an offer and the contractor carries that subtrade in the prime bid.
- When a contractor carries a subcontractor by listing its name and incorporating its price in the prime bid, Contract A will oblige the contractor to communicate acceptance of Contract B to that subtrade if the prime bid is accepted by the owner, and the owner does not reject that particular subtrade.
- If the contractor carries a subtrade's price but does not name the subtrade, the terms of Contract A may still oblige a contractor to communicate acceptance of Contract B to the subtrade if the owner accepts the prime bid without objection to that subtrade (especially since naming the subtrade only confirms an event that has already occurred, namely acceptance of that subtrade's price).
- Contract A between contractor and subtrade ends if the owner rejects the general's bid or does not accept it within the period of the bid's irrevocability.
- The contractor is not obliged to nominate the lowest subcontract bidder.
- Previously merely anti-social, bid shopping is now a breach of Contract A.
- The offer of a subcontractor remains open and irrevocable for a "reasonable" time after the expiry of the period of irrevocability in the prime bid.

# Chapter 5

# SUPPLIER BIDS

*For a time, it seemed that the decision of the Nova Scotia Supreme Court in the following case would force architects and engineers to make their roof specifications water- and airtight enough to resist legal scrutiny as well as the elements. The decision has been reversed on appeal. So back to the old order of things — until the next challenge.*

### ARROW CONSTRUCTION PRODUCTS LTD. v. NOVA SCOTIA (ATTORNEY GENERAL)

Nova Scotia Court of Appeal; April 1996

In February 1993, the Province of Nova Scotia called for bids for the repair of a large roof on a maintenance facility outside Halifax and specified that the roofing material should be blue PVC. Arrow Construction Products Ltd., the principal supplier of EPDM membranes in the Atlantic area, lobbied the Province to permit bids based on EPDM.

The Province issued an addendum to allow white or gray PVC or white EPDM roofs. The base bid for the project was to be for blue PVC, but, "after determination of preferred bidder", alternatives and corresponding bid price adjustments would also be considered.

Arrow did not bid, but submitted prices for the supply of EPDM to several contractors who did.

J.W. Lindsay Enterprises Ltd. submitted the lowest base bid of $373,465. It indicated that it would deduct $29,434 from the base bid if the owner accepted a white PVC membrane instead of blue, or it would deduct an additional $1,800 for EPDM. The Province opted for white PVC in spite of the price difference. After researching the materials, its architect had come to the conclusion that PVC allows better seams under winter conditions.

Lindsay was advised that it would be awarded the contract on the basis of its base bid. It was also issued a change order for the installation of white PVC, which resulted in a reduction of $29,434 for a revised contract price of $344,031.

Arrow sued the Province, claiming damages for loss of profit caused by negligent misrepresentation, express or implied, that Arrow would receive fair treatment in the assessment of EPDM.

Expert evidence at trial convinced the judge that PVC and EPDM were equal materials. He concluded that the contract should, therefore, have been awarded to the lowest *overall* bid rather than the best *base price* bid as the owner had done.

He agreed with Arrow that the Instructions to Bidders and the addendum constituted a representation that the Province would give "fair consideration" to the use of EPDM. It did not do so. Arrow would have earned a profit if the contract had been awarded to a bidder who used EPDM. When the Province chose PVC, Arrow saw its efforts wasted. It relied on the Province's misrepresentation and, as a consequence, suffered a loss.

The trial judge also found that the Province had a "hidden preference" for PVC. This could not be supported by the usual privilege clause in the bid documents that reserves to the owner the right to accept or reject any or all bids.

Arrow did not quantify its loss by any specific evidence relating to the cost of its bid or anticipated profits. The trial judge, nevertheless, awarded Arrow general damages of $25,000 and legal costs. The Province appealed.

Underlying the trial judge's finding that the Province was liable was the judge's finding that EPDM and PVC were equal, a conclusion based on a contest between the experts and other witnesses as to the comparative merits of the two materials.

"In my opinion, this was an enquiry on which [the trial judge] should not have become sidetracked", said Justice Chipman writing for a unanimous Court of Appeal. Nowhere in the bid or contract documents did the owner treat PVC and EPDM as equal. It could have, but it did not choose to do so. The owner's preference was clear from the start.

When using the base bid approach, the specifier can choose a material which it has carefully evaluated. It can still obtain competitive bids by using one of two common methods:

- pre-selecting the products of other manufacturers as "approved equals"; or
- listing "approved alternatives" and then requiring bidders to indicate the addition or deduction from the base bid price.

The Province chose the second alternative, said Justice Chipman, but the trial judge imposed the first alternative on the owner and so amended the bid documents. Justice Chipman was very critical of this approach:

The trial judge was thus rewriting the tender call on the basis of his engineering judgment that EPDM and PVC were equals. With respect, I know of no authority for such judicial review of engineering decisions in the process of awarding a tender and there is no place for such a process which would introduce chaos into the courts and into the construction trades.

The owner is entitled to make a choice of building materials free from judicial review.

In addition to the irrelevant inquiry into the merits of the roofing materials, said Justice Chipman, the trial judge entered into an equally irrelevant inquiry into the operation of the bidding system that the Province had in place. The judge found many faults in the system. None of this, however, was relevant to the question before the judge: did the Province make a negligent misrepresentation to Arrow?

The Court of Appeal found no negligent misrepresentation. The owner only promised to consider EPDM. After the bids were in, the owner did just that; then it decided to use PVC.

The requirements for a successful claim of negligent misrepresentation are based on the classic *Hedley Byrne* principles, which were clarified by the Supreme Court of Canada in *Queen v. Cognos* as follows:

1. There must be a duty of care based on a "special relationship" of closeness or proximity between the person making the representation (the representor) and the person to whom the representation is made (the representee).
2. The representation in question must be untrue, inaccurate, or misleading — in other words, it must be a misrepresentation.
3. The representor must have acted negligently in making the misrepresentation.
4. The representee must have relied, in a reasonable manner, on the negligent misrepresentation.
5. The representee must have suffered damages as a result of his or her reliance on the misrepresentation.

Arrow clearly failed to pass the test of the second requirement: the owner made no representations to Arrow that were untrue, inaccurate or misleading.

Arrow also failed the fourth requirement. How could Arrow reasonably contend that it relied on an untrue, inaccurate or misleading statement that EPDM would be chosen based on considerations of "fairness"? How could Arrow, in view of the terms of the bid documents, have relied in a reasonable manner on the addendum as a representation that the owner would award the contract to the lowest *overall* bidder?

Besides, the *Hedley Byrne* principle deals with representations, not promises. There is a big difference: a representation is a statement relating to some existing fact or past event; a promise is a statement of intention to do something in the future. Only representations of existing facts, and not those relating to future occurrences, can give rise to actionable negligence.

The Province's promise to consider alternatives to PVC was a promise dealing with a future action. "It is a play on words to say that it falsely stated its real intention respecting a preference for PVC or EPDM", said Justice Chipman.

The trial judge also found that the owner had a hidden preference for PVC. Yet the specifications that went out as part of the bid documents contained six pages dealing with the supply and installation of PVC and nothing on EPDM. While the addendum indicated a willingness to consider alternatives, it did not constitute a repeal of this clearly expressed preference.

The precedent-setting cases on which the trial judge relied established that, subject to the provisions in the bid documents, an owner (and especially a public-sector owner) cannot use the privilege clause to impose a *secret* preference for a particular bidder or class of bidders (typically, local contractors), or for a particular material. However, there are commercial considerations which the owner can, in the public interest, take into account.

The Court of Appeal was satisfied that there was no lack of fairness in the Province's bidding process, contrary to the findings of the trial judge. The trial judge's conclusion that the owner failed to act with fairness to Arrow, said Justice Chipman, arose from his "unwarranted" inquiry into the merits of PVC and EPDM:

> By converting alternates into equals, the trial judge set the stage for his findings that there was misrepresentation, that the preference for PVC was a hidden preference, and that there was a lack of fairness.

The Construction Association of Nova Scotia, acting as intervenor in the action, submitted that the owner had a duty to evaluate alternative products fairly or reasonably. Justice Chipman disagreed:

> I think it is the prerogative of an owner to make its own judgment on which alternative it chooses... Such a submission would only warrant consideration if the documents in the tendering process imposed such a duty.
>
> To accept the intervenor's proposition would run the risk of rendering every engineering choice by an owner subject to judicial review, every disappointed bidder looking for ways to challenge an award of a tender in the courts.
>
> The argument of the intervenor confuses the right of choice of an owner with improper practices of calling for tenders with hidden preferences.

The court allowed the appeal and set aside the decision of the trial judge. It also awarded the Province the costs of the trial and of the appeal.

*What happens when a supplier makes a mistake in the price offered to a subcontractor which then uses this price in its bid to the general contractor?*

*The next case examines the ramifications in detail, and also offers a stinging condemnation of bid shopping in general.*

## WESTERN PLUMBING AND HEATING LTD. v. INDUSTRIAL BOILER-TECH INC.

Supreme Court of Nova Scotia; September 1999

Roscoe Construction Ltd., a general contractor, was preparing its bid on a laboratory project in Nova Scotia. The bid documents stipulated that the bids were to remain open for acceptance for 30 days after bid closing on January 15, 1997. That date was later extended to January 17.

Western Plumbing and Heating Ltd. was compiling prices in order to submit a subtrade bid to Roscoe. As part of the process, Western obtained a quotation for two boilers from Industrial Boiler-Tech Inc. The price offered was $52,161 for both boilers. The bid was neither signed nor was it under seal.

Industrial's offer made no reference to irrevocability, and the supplier had no knowledge of the terms of the bid call as they applied to prime contractors.

Western received two other bids for the boilers. Industrial's bid was the lowest; the other two were significantly higher: one for over $74,000 and the other for almost $95,000. Western carried Industrial's quote in its bid to Roscoe. Finally, after a delay in the award of the prime contract, Roscoe awarded the subcontract to Western on February 28.

On March 4, Western contacted Industrial. Its purchaser told the manager of Industrial that the price quoted was too high, but, if Industrial would reduce the price to $50,000, it could have the contract. The purchaser explained that he had another bid in that amount from another supplier.

Industrial then reviewed its price and discovered that it had made a serious mistake. Its price should have been $74,561. It faxed Western the revised price and stated that it would not honour the original price of $52,161. It also suggested that if Western could buy the boilers elsewhere for $50,000, it should do so before that price went up.

On March 7, Western sent a purchase order to Industrial for the boilers for $52,161. Industrial refused to deliver the boilers, and Western purchased them from the next lowest bidder for $66,408. It then sued Industrial for the difference between its original offer and the price paid, *i.e.*, $14,247.

Justice Gruchy of the Supreme Court of Nova Scotia reviewed the factual and legal issues one by one.

What was the effect of Industrial's mistake? Industrial submitted that, where a subcontractor knew that a supplier had made a mistake in a bid, he could not accept such a mistaken bid and enforce it. It cited, among other authorities, the *Belle River* decision summarized in this book and, of course, *Ron Engineering*. However, Justice Gruchy said:

> I am unable to conclude that any of these cases stand for the proposition that a contractor is under a legal obligation to inquire into or question the validity of a price on the mere suspicion that it appears to be low.

He then quoted extensively from the decision in *Gloge*. In particular, he noted two points made by the court.

First, when a general contractor invites bids from subcontractors, it wants to hold those bids open for acceptance until the general contractor knows whether its bid has been accepted by the owner.

Second, there is a great difference between the consequences of a mistake in a bid by a general contractor to the owner, and that of a subcontractor to the general contractor:

- When the general contractor makes a mistake in its bid, it is the author of the mistake that will suffer if the contractor is held to the contract.
- If the subcontractor makes a mistake, and the general contractor is held to its contract with the owner while the subcontractor is relieved of its commitment, then an innocent party will suffer for the mistake while the perpetrator of the error escapes all liability.

But following *Ron Engineering* and *Gloge*, the subcontractor cannot escape liability. The trial judge in *Gloge* took from *Ron* a simple proposition: If, at the time the bids are opened and Contract A comes into existence, and the recipient of the mistaken bid does not know of the mistake, the court does not need to even consider the law of mistake as it was applied in *Belle River*.

Justice Gruchy agreed and rejected this line of defence. Western had reason to be suspicious of the very low price, but that did not mean it actually knew that Industrial had made a mistake. Western's witness stated that "prices can change overnight" and "there is always time for one more phone call". However, the statements would come back to haunt Western.

Was Industrial's bid revocable? The bid contained no time limit for acceptance. Industrial was not aware of the context of the bid call issued to general contractors, or of the Instructions to Bidders, or of the time extensions and delays in the awarding of the contract.

The judge decided that, clearly, a bid by a subcontractor should be open for acceptance for a reasonable period of time. He accepted that the 30 days for acceptance of the prime contract was reasonable and that the period of acceptance of subcontracts would have to be within a reasonable period after that closing date.

Industrial could reasonably have concluded that 30 days would be the time for acceptance of the prime contract, and its bid would have to be open for acceptance for a reasonable time after that. The court found this to be an implied term of Industrial's bid. But Western did not accept the bid within that period. Sixteen days elapsed before Western notified Industrial of its acceptance. Justice Gruchy found this period unreasonable and concluded that it was open for Industrial to revoke its bid — and it did do so.

The judge came to this conclusion, bearing in mind Western's remark that "prices can change daily". In the circumstances, it would be unreasonable to assume that Industrial would hold its offer open for acceptance indefinitely. Furthermore, Industrial was not involved in the formation of Contract A and, accordingly, could not be bound by the terms and conditions of the bid call.

Was the revocation valid? When Western asked Industrial to reduce its price, Industrial revoked the original bid and offered the boilers at the higher price. This offer was not accepted.

Western's purchase order was ineffective, since it was forwarded on March 7, after Industrial's revocation on March 4. The court concluded that Industrial's original bid was not validly accepted but that it was validly revoked before Western's purported acceptance.

Did Industrial's action in advising Western that it would supply boilers at the price quoted and then failing to do so amount to the tort of negligent misrepresentation? Western pleaded that it was. The judge agreed, but that did not help the subcontractor. Western had the opportunity, in reliance on the misrepresentation, either to accept the bid within a reasonable time or to ask that it be extended. It did neither.

The negligent offer expired, and then Western, in attempting to get a better price from Industrial, made its own deliberate misrepresentation that it could buy the boilers for $50,000. Industrial relied on that when it revised its price and suggested to Western that it should snap up the boilers at the price quoted.

Finally, the subject of bid shopping came up. It was unnecessary for the judge to address this, but he felt "some compulsion to express my views and conclusions on it". This he did, and he did not mince his words.

At trial, several witnesses testified that contractors, after winning the prime contract, routinely ask subcontractors to lower their bids. Subcontractors and suppliers almost never refuse to bargain, and they accept that practice.

Other witnesses described this practice as unethical but would not call it bid shopping. True bid shopping is when a contractor obtains a price from a subcontractor and then uses that price as a bargaining chip to obtain better prices from other suppliers or subcontractors.

The judge found that the process followed by Western amounted to bid shopping and bordered on deceit. He commented:

> ... the contractors and subcontractors have turned the tendering process into a sham. The assertion that the practice described to me is not bid shopping appears to be sophistry and hypocritical. The practice encourages subcontractors and suppliers to put in inflated bids, keeping in mind that ultimately the contractor will attempt to force them to lower their prices. The lowered prices will result in a profit to the contractor, but with no saving passed on to the owner. The practice will ultimately defeat (and arguably has done so) the value of a bona fide tendering process. In effect, Western has by this action [law suit] requested the court to validate the bargaining process as I have described it. I will not do that.

Justice Gruchy dismissed Western's action with costs.

---

Finally, a trial judge has observed that the emperor is not wearing any clothes. Unswayed by the evidence to the contrary, Justice Gruchy found that the attempt to extract $2,000 from Western Plumbing in order to secure the contract was as much "buck naked" bid shopping as using another supplier's price to accomplish the same goal.

The judge's comments on bid shopping may be of limited value since the trial judge also found that this was not a bid in the Contract A/Contract B sense. Is there anything wrong with negotiations about price in ordinary day-to-day offer/acceptance dialogue? Probably not.

The trial judge's discussion on bid shopping was timely and pertinent but was not the central issue in the case. Nevertheless, bid shopping is beginning to take on a breach of Contract A pallor — at least under the bright lights in court.

The issue was whether or not this particular supplier fell within the Contract A/Contract B structure that governed the activities of the owner, contractors and subcontractors in the bidding process. The judge decided that the supplier fell outside:

- *first*, because it did not know about it; and
- *second*, because the price it submitted for the boilers was a simple, revocable commercial offer having none of the indicia of a bid such as irrevocability, bid security and some conformance to the bid documents of the prime contract.

*From the perspective of the supplier to a potential subtrade, the lesson in this case is to include language in the offer permitting the offer to be withdrawn at any time and providing that it expires if not accepted by a specified date. The ability of the supplier to form its offer this way depends on the requirements laid down by the subtrade.*

*For subtrades, the lesson is to communicate the purpose of seeking the offer and to secure a period of irrevocability which, at least, matches its own obligation in its bid to the contractor.*

## SUMMARY

The following basic lessons can be drawn from the cases summarized in this section:

- Supplier pricing to a contractor or subcontractor may or may not be covered by the principles of *Ron Engineering* — it depends on the documents that solicit the price.
- The owner is entitled to make a choice of building materials free from judicial review; owner choice, when specified, must not be confused with improper bid practices such as hidden preferences.
- A contractor or subcontractor is not under a legal obligation to investigate whether a price submitted to it was submitted by mistake if it merely suspects that the price appears to be too low.
- It is an implied term of a supplier's bid (if it is a bid at all) that its price will be open for acceptance during the same period as that of the prime contract, and for a reasonable time after it.
- A supplier which advises the contractor or subcontractor that it will supply a given product at a given price may be liable for misrepresentation if it fails to keep its commitment.

# Chapter 6

# MISTAKE IN BID

*The Ron Engineering decision changed the foundations of the law of competitive bidding. The most dramatic impact was felt in the treatment of mistaken bids, starting with the Ron Engineering case itself and followed soon after by the Northern Construction case.*

*The next case applies the Ron Engineering principles to the bidding relationship between a contractor and a subcontractor. In this case, it is the subcontractor that has made a mistake in its bid to the general contractor.*

### GLOGE HEATING & PLUMBING LTD. v. NORTHERN CONST. COMPANY LTD.

Court of Appeal of Alberta; January 1986

Gloge Heating & Plumbing was one of a number of mechanical subcontractors that had prepared a bid for submission to each of the general contractors bidding on the construction of the Edmonton International Airport.

Gloge telephoned its bid to Northern Construction Company and to other general contractors minutes before bids closed in order to avoid "bid shopping" by general contractors. Northern completed its own bid and named Gloge for the mechanical work, as its bid was the lowest. Shortly afterwards, Gloge realized that the bid contained a serious mathematical error that resulted in a substantial shortfall.

Northern was the lowest general bidder. Gloge advised the general contractor of the error. Northern, in turn, advised the owner before its bid had been formally accepted, but the owner would not permit Northern to adjust its bid. Northern was awarded the contract. Gloge refused to perform the mechanical subcontract.

Northern made alternative arrangements with another subcontractor at an increase of $341,299 in the subcontract price and sued Gloge for that amount.

Gloge defended by putting forward two arguments:

- Northern had failed in its duty to alert Gloge that its bid was so low it had to be erroneous.

- Northern could not claim to have accepted Gloge's bid after it knew that Gloge had made a mathematical error.

The court disposed of these arguments summarily, holding that Gloge, by delaying the submission of its bid to the last minute, had deprived Northern of any real opportunity to analyze the bid or compare it with others. When Northern learned of the error, it was too late to assist Gloge without the owner's consent, and that consent was denied.

Gloge also tried the traditional principle that it was free to withdraw its bid at any time before Northern accepted it, and that it had done so — see *McMaster University v. Wilchar Construction*. This argument was the real issue of the trial.

According to the analysis in *Ron Engineering*, Northern made an offer to Gloge to submit a bid for the mechanical subcontract. Gloge accepted this offer by submitting the bid. Thus, a valid contract was made (Contract A). Gloge knew that Northern would rely on its bid and that Northern's bid would be irrevocable. This was enough for the court to find that Gloge's bid to Northern was also irrevocable.

The court also held that it was normal and standard practice for general contractors to accept last minute telephone bids from subcontractors, and that it was understood and accepted by the construction industry that, while such bids could be withdrawn prior to the close of bidding, they would remain irrevocable after bids closed for the same term that the general contractors' bids to the owner were irrevocable. Accordingly, Gloge was obligated to perform the work when Northern awarded it the subcontract and was, therefore, liable for failing to do so.

---

*Although the case law between owner and contractor, and between contractor and subcontractor is of the same family, the predicaments of the owner and the contractor are quite different. If the subcontractor has made a mistake which comes to light after the prime bid has been submitted, the contractor has committed itself to the owner and relied on the subtrade price. The contractor will absorb the loss if the trade withdraws and the owner will not give the contractor relief. The contractor has a real loss in having relied on the subtrade.*

*The owner in the Ron Engineering case was in a very different position. It had not taken any step in relying on the contractor's mistaken bid. If the contractor had not been obliged to proceed, the owner would have been deprived of an offer which the contractor had never intended to make.*

*The perceived evil that Ron Engineering was attempting to address was that an irrevocable bid supported by bid security was not an irrevocable bid at all if the declaration of "mistake" allowed the contractor to avoid the consequences* of its error. Even without Ron Engineering, the outcome in Gloge would probably have been the same because Northern, to its detriment, relied on the price which Gloge provided. The measure of its detriment was $341,000, which was the spread between the Gloge price and the price of the replacement trade it was obliged to hire to carry out Gloge's work.

## TOWN OF VAUGHAN v. ALTA SURETY COMPANY

Supreme Court of Ontario; May 1990

In 1986, the Town of Vaughan, Ontario, called for bids on a construction project. The bids were to be under seal, and each was to be accompanied by a bid bond of $250,000. Acme Building and Construction Limited submitted a bid in the amount of $5,030,000. Included with the bid was a bid bond issued by Alta Surety Company.

Acme's estimator was present when the bids were opened on October 14, 1986. He later testified:

> My stomach sank when the bids were opened. We were the lowest by $250,000 and the other bids were all within dollars one of the other.

He checked his company's bid later that evening but found no mistake. Only later was he alerted to the fact that he had failed to include the cost of the sprinkler system at $109,200.

The following day, the vice-president of Acme called the project architect and then wrote a letter:

> We also confirm our telephone conversation today... wherein we informed you of an error in our tender amounting to about one hundred and ten thousand dollars ($110,000.00). We inadvertently left out the sprinkler work, Section #15D of the Specifications... We were unaware that the sprinkler work had been tendered separately through the depository and as we did not receive any bid for the sprinklers, we were not aware of the omission....
>
> This error is certainly demonstrable and it can be verified by checking with the bid depository that we did not receive any tender for the sprinkler work. Accordingly we request that our tender be adjusted by the full amount of the sprinkler tender as submitted by Wray Sprinkler Company Limited.

Vaughan's director of property recommended to council that a contract be awarded to Acme for $5,030,000 — the amount of its allegedly mistaken bid. Vaughan council accepted the recommendation at its meeting of October 20.

Nine days later, Acme advised the owner that it was not prepared to sign the contract unless the bid was adjusted by the amount of the error. It pointed out that if that cost was included, Acme's bid would still be significantly lower than that of the second lowest bidder.

The owner, however, requested that Acme provide the architect with a performance bond and a labour and materials payment bond within two weeks of the date of notification of contract award, that is, by November 5, and that it execute the construction contract as provided by the bid documents.

A subsequent letter informed the contractor that its request for an increase in the contract amount was unacceptable. The owner stated that it intended to call on the contractor's bid bond and award the contract to the second lowest bidder.

On December 1, Acme capitulated. It informed the owner that it was prepared to execute the contract in accordance with its bid. But it was too late. On December 8, the council accepted the recommendation of its director of property to accept the bid of the second lowest bidder. Soon after, Vaughan's solicitor called on Alta Surety Company to honour its bid bond. When Alta did not do so, Vaughan launched its lawsuit against both Alta and the general contractor.

The court found that Acme had breached Contract A. Acme stated all along that it would not sign the construction contract unless the owner added $109,200 for the sprinkler system to the bid. The letter of December 1 was the only time that Acme said otherwise.

Did that letter constitute a timely legal cure of the breach of contract? Counsel for Acme referred to it as the "letter of capitulation". Counsel for the owner called it "death bed repentance". For the purposes of the litigation, said the judge, it does not matter whether it is one or the other.

Acme was notified verbally on October 22 that it had been awarded the contract, and, on November 4, it was notified by letter. Acme did nothing until it knew that the second lowest bidder was going to get the award. The two-week time period in the bid contract allowing for the delivery of the bonds started to run from October 22.

Giving Acme the benefit of the doubt that the time period did not start to run before November 14, the two-week deadline would still have expired on November 28, well before December 1. Thus the time within which to cure the breach had expired before Acme attempted to do so.

Counsel for Acme argued that, had the owner mitigated its damages, there wouldn't be any:

> ... if the owner had added the cost of the sprinkler system to the bid and awarded the contract to Acme, the price would have been $139,800 less than the second lowest bid.

In other words, if the owner had awarded the contract to Acme at the bid price of $5,030,000, when Acme finally agreed to it, the owner would have suffered no damages.

The court disposed of this argument by quoting from the judgment of the Alberta Court of Appeal in *Calgary v. Northern Construction Company*. In that case, the contractor similarly argued that the owner should have accepted the contractor's bid plus the amount of the clerical error the contractor had made because this would have mitigated its damages. The passage from the decision of the Court of Appeal in that case was to the point (emphasis added):

> Undoubtedly [Calgary] had the duty to mitigate its damages but to accept the argument of the contractor would be to *change the tendering system to that of an auction*. To accept the submission of the contractor would allow any contractor who made a low bid to refuse the contract but to offer to do the work for less than the second low bidder, and then argue the city must accept such offer in mitigation of its damages. The city was under no such duty and the contractor has not proven any failure of the city to mitigate.

The court found that the Town of Vaughan was entitled to judgment against Acme and Alta in the sum of $249,000, the penalty of the bid bond. The trial decision was affirmed in the Court of Appeal, and leave to appeal to the Supreme Court of Canada was refused.

---

*In contract law, offer and acceptance must correspond exactly (for Contract A, substantially). Suppose A offers to sell B a car for $1,000. If B says, "I agree to buy the car but will only pay $950", there is no contract. In general, if the offeree makes a counteroffer, the offeree is deemed to have rejected the original offer and it cannot subsequently change its mind and accept it.*

*In the following case, the subcontractor tried this principle to save itself from the consequences of a mistaken bid.*

## FOREST CONTRACT MANAGEMENT v. C&M ELEVATOR LTD.

Court of Queen's Bench of Alberta; October 1988

In July 1986, Forest Contract Management Ltd. requested telephone quotations from several subcontractors to supply and install two hydraulic elevators for an apartment building in Edmonton. C&M Elevator Ltd. submitted its bid in the amount of $94,720.

Forest in turn submitted the owner its general bid and was awarded the contract. C&M confirmed its telephone quote by letter. Some of the conditions contained in the letter were not acceptable to Forest and, after negotiations, were changed. The price was also renegotiated down to $80,312.

The construction contract between the owner and Forest was signed in December, and, the following January, C&M received the subcontract documents. C&M then requested drawings and specifications for the elevators, obtained a cost from its supplier, Northern Elevator, helped Forest locate the holes for the elevator pistons and took other steps to fulfill its obligations under the subcontract.

However, on March 30, C&M called Forest and informed it that Northern Elevator had missed one elevator in its estimate and that C&M could not perform the job for the price quoted. This announcement was followed by a letter from the subcontractor's president, stating:

> We regret to inform you that, due to circumstances beyond our control, we are forced to withdraw from this project at this time. We are therefore returning your contract to you, unsigned.

Forest re-tendered the contract, and Dover Elevator was awarded the job at a cost of $99,000. The contractor then sued C&M for breach of contract and claimed damages in the sum of $18,388.

The subcontractor had returned the formal contract to Forest without signing it. However, relying on the decision in *Ron Engineering*, Justice Montgomery found that, nevertheless, there was a contract — Contract A — between Forest and C&M. The contract that C&M received and returned unsigned was Contract B, the second component of the bidding process.

Counsel for C&M submitted that Contract B should reasonably reflect the terms of Contract A. He argued that the change in the contract price and the various other changes constituted a *counteroffer* and that C&M was not bound by the contract until it was formally executed by both parties. As this never happened, Contract B never came into force.

The court found the changes insignificant. In the opinion of the court, C&M was obliged to review Contract B before it started work on the job, and if the terms did not match Contract A, it should have informed Forest.

The subcontractor never stated that it would not perform Contract B. On the contrary, it operated between early January and March 30 as though Contract B was in place. Such performance by C&M brought Contract B into force:

> In this case there was performance by C&M after receipt of Contract B which in my judgment takes Contract B out of the requirement established in the *Calgary [v. Northern Construction]* case where it was stated that Contract B must be executed by both parties before it comes into force.

The judge went on to explicitly formulate the rule:

> Once a subcontractor commences to perform Contract B, it is bound by the terms of Contract B. At that stage, execution of Contract B by the parties is no longer a requirement. If the terms of Contract B do not conform with Contract A, the sub-

contractor should notify the general contractor immediately and not perform until all of the terms of Contract B are settled.

Forest was awarded damages in the amount of $18,388, the difference between the Dover price of $99,000 and the price in its Contract B with C&M.

*Several cases summarized so far have shown that a substantial error in a bid can be a problem both for the owner and for the bidder. The following case shows that a seemingly **trivial** error can also be fatal to a contractor's bid.*

## VACHON CONSTRUCTION LTD. v. CARIBOO (REGIONAL DISTRICT)

Court of Appeal for British Columbia; June 1996

Can-form Construction Limited was the lowest bidder on a bid call by the Cariboo Region in British Columbia. Vachon Construction Ltd. was the second lowest bidder.

Following the opening of the bids, it was discovered that there was a clerical error in Can-form's bid. The bid amount was expressed in *words* as four hundred and eighty-eight thousand four hundred and fifty dollars ($488,450), but, immediately following this text, in brackets, there appeared a larger *numerical* amount: $492,450.

An authorized representative of Cariboo asked Can-form to indicate which of the two numbers was the bid price and was advised that the lower figure represented the bid. The higher figure was then crossed off and the document initialled.

The remaining two bids, including Vachon's, were then opened, and both were higher than either of the two amounts in Can-form's bid.

Vachon wrote a letter to Cariboo, complaining about the Can-form bid and asking that the contract be awarded to Vachon, the next lowest bidder. Cariboo, nevertheless, awarded the contract to Can-form.

Vachon sued. It claimed that Can-form's bid was invalid, and that in accepting that bid, Cariboo had:

- breached Contract A, which had been created with Vachon when it submitted its bid; and
- breached the duty of fairness it owed to Vachon for having submitted the lowest *valid* bid in compliance with the bid documents.

Vachon's claim was heard in chambers. The chambers judge looked at the Instructions to Bidders. Among other provisions, they contained the following:

- Amendments to a submitted offer will be permitted if received in writing prior to Tender closing and if endorsed by the same party or parties who signed and sealed the offer.
- Tenders that are unsigned, improperly signed or sealed, conditional, illegible, obscure, contain arithmetical errors, erasures, alterations, or irregularities of any kind *may* be considered informal.
- Tender Forms and enclosures which are improperly prepared *may* be declared informal. (emphasis added)
- The owner reserves the right to accept or reject any or all offers.

The chambers judge decided that when Can-form's representative at the opening of the bids picked one of the two amounts as the right one, that was not an amendment to the bid as contemplated by the Instructions to Bidders, but an "irregularity". Cariboo was, therefore, entitled to exercise its discretion and permit the bidder to correct the price discrepancy.

Vachon appealed. The Court of Appeal first had to decide whether Can-form's bid was valid as submitted. "In my view it was not", decided Justice Finch, and continued:

> Price is an essential element of a bid in response to an invitation to tender, and an offer which is uncertain as to price cannot form the basis of a binding contractual relationship.

In *Ron Engineering*, the Supreme Court considered the possibility of bids being defective for some reason and the consequences that would follow. Justice Estey said:

> There may well be, as I have indicated, a situation in the contemplation of the law where a form of tender was so lacking as not to amount in law to a tender....

Justice Finch applied this statement to the case before him and found that Can-form's bid, as submitted, could not give rise to Contract A because it did not clearly state the bid amount. This made Can-form's bid invalid.

The next question was whether Can-form's invalid bid could be corrected by choosing between the two alternatives.

Justice Finch agreed with the chambers judge that the owner had a "discretion" but, in his view, it was limited to treating bids as either "informal" or not. The owner was not entitled to alter or correct an irregularity in the bid after the close of bidding, nor was it entitled to attempt to render valid, after opening, a bid that was invalid when it was submitted. The question created by the ambiguity of Can-form's bid was whether the bid was valid, not whether it was irregular.

The Instructions to Bidders quoted above confirmed this view. They stipulated that amendments to a bid would be permitted only "if received in writing *prior* to Tender closing". This showed an intention not to permit amendments *after* the close of bids.

"Such an interpretation is consistent with the language used, with common sense, and with the goal of protecting the integrity of the tendering process", said Justice Finch and concluded that it was not open to Can-form to alter or correct the error in its bid concerning price, either with or without the agreement of the owner.

Where did this leave Vachon as the next lowest bidder? Cariboo's position was that the privilege clause which gave the owner "the right to accept or reject any or all offers" was a complete answer to Vachon's claim.

Contract A arose between Vachon and Cariboo when Vachon submitted a valid bid. However, the formation of Contract A did not give Vachon the right to enter into Contract B, that is, the construction contract. So what rights did Vachon acquire under Contract A?

In spite of the privilege clause, decided the court, the owner still owes a duty of fairness to bidders in accordance with the decision of the Federal Court of Appeal in *Best Cleaners.* In that case, Justice Mahoney said:

> [The owner's] obligation under contract A was not to award a contract except in accordance with the terms of the tender call. The stipulation that the lowest or any tender need not be accepted does not alter that.

The reasoning of *Best Cleaners* was picked up by the B.C. Court of Appeal in *Chinook Aggregates.* Abbotsford had an undisclosed policy of preferring bids from local contractors whose bids were within 10% of the lowest bid. In spite of the privilege clause in the bid documents, Abbotsford was held to be "in breach of a duty to treat all bidders fairly and not to give any of them an unfair advantage over the others".

The Ontario Court of Appeal did not follow *Chinook Aggregates* in *Acme Building & Construction Ltd. v. Newcastle.* The court held that even if industry custom and usage mandated the acceptance of the lowest bid, that could not prevail over the clear and express language of the privilege clause.

The Ontario court, however, did not reject the argument that there is a duty of fairness to all bidders. It suggested, as well, that industry usage and custom *can* place implied terms in contracts where these are not inconsistent with the written contract. Justice Finch concluded that his view was not inconsistent with the decision of the Ontario Court of Appeal.

Although the facts in *Chinook Aggregates* were different from the facts in the *Vachon* case, there was a common element. In both cases the owner gave an advantage to the successful bidder.

In *Chinook Aggregates,* the advantage was a preference for local contractors that was not disclosed in the Instructions to Bidders and was unknown to the bidders. The advantage to Can-form was an opportunity to amend its bid in a way that was not in accordance with the Instructions to Bidders. In *Chinook Aggregates*, the court held that the owner breached the duty of fairness. So did Cariboo.

Vachon submitted its bid with the expectation that all bids would be considered according to the rules laid down in the Instructions to Bidders. It would be unfair to allow Cariboo to act in breach of those rules and then to rely on the privilege clause to escape the consequences, concluded Justice Finch as he allowed Vachon's appeal.

The amount of damages to which Vachon was entitled by reason of Cariboo's breach of Contract A will be decided at another trial.

Justice Williams concurred with Justice Finch (as did Justice Prowse, making the decision unanimous) and added his own comments, which explain with particular clarity why the courts treat mistaken bids with such rigour.

The decision of the chambers judge, said Justice Williams, was based on common sense, but, unfortunately, common sense does not always accord with common law.

The discrepancy in the amount of Can-form's bid was not a large amount. It was also very clearly an error. Even the highest amount in Can-form's bid was lower than the next lowest bidder. It would seem that the chambers judge was correct in putting substance over form and holding that the owner was justified in exercising its power to consider the obvious mistake as an "irregularity" and that the court should allow that irregularity to be corrected.

The owner would, thus, achieve the lowest price for the work, and Can-form would succeed as the lowest bidder. There would be no change in the order of bids, since Can-form was the lowest bidder in any case.

On the other hand, the bidding process is, and must always be, carefully controlled, since the opportunity for abuse or distortion is ever present. It must be, and be seen to be, fair to all bidders. For that reason, the law has applied strict rules for any alteration in the process by both bidder and owner.

In Can-form's case, while the mistake was obvious, it was not obvious whether the lower or higher figure was the actual price intended. Price is a critical part of the bid, and a mistake in the price must be considered more than an irregularity.

Was the bid valid? That question should have been answered at the time the bid was submitted. If the defect in the bid was more than an irregularity, then the bid was simply not valid as a matter of law. If the bid was invalid, it was irrelevant whether either of Can-form's two bid amounts were lower than the next lowest bid or not.

If the owner is permitted to change a bid to correct an irregularity, when should that be allowed? Should it be allowed before the other bids are opened, as in this case? What if the irregularity appeared when the second bid was opened? What if the amount of the second bid was in *between* the worded and the numerical amounts of the first bid opened? The ramifications of correcting a bid are endless, and this explains why any changes to the bid must be made prior to bid closing in order to preserve the fairness of the process.

For those reasons, the decision of the chambers judge could not be allowed to stand.

---

*The scale of the cost to the bidder of making a mistake is greater in the next case than in any of the previous ones — more than 40% of the bid! Quite apart from the legal issues, the case raises the question of the cost to society of such an outcome. A penalty for this magnitude of an error can easily wipe out a contractor. Is this kind of sacrifice really justified simply in order to preserve the sanctity of the bidding process?*

## VIC VAN ISLE CONSTRUCTION LTD. v. SCHOOL DISTRICT NO. 23

Court of Appeal for British Columbia; March 1997

Over the years, contractors have learned to live with the fact that if they make a mistake in their bids, the law gives them no way out, and they must suffer the consequences — the sanctity of the process of competitive bidding.

Vic Van Isle Construction Ltd. learned this lesson on a project in Williams Lake, B.C. After the bids closed, Vic Van Isle discovered that it had made a $100,000 error in its calculations. It tried to withdraw, but was informed that the law did not permit such relief. The bidding contract, or Contract A, was binding. The owner accepted its bid, and Vic Van Isle was compelled to complete the project for the mistaken amount.

Soon after, the owner's architect made an error on the project that seemed to give Vic Van Isle a chance to save some money due to the rigidity of Contract A. However, it was not to be.

In February 1993, Vic Van Isle was the lowest bidder on a school project for the School District No. 23 (Central Okanagan) near Kelowna, B.C. The problem leading to litigation arose because provincial legislation obliged contractors to follow certain fair wage requirements, but there were two sets of incompatible requirements in force in the province at the time.

In 1976, the B.C. Legislature enacted the *Wage (Public Construction) Act*, which prescribed that on every public project, the workers must be paid fair wages, defined in terms of wages and benefits usually paid to workers in the geographical area of the project.

In 1992, the B.C. Minister of Labour issued the *Fair Wage Order*, which, without reference to any statutory authority, imposed arbitrary provincewide minimum hourly rates and benefits to be paid to workers on public construction projects. The rates were significantly higher than those contemplated by the 1976 Act.

Vic Van Isle's fixed-price bid was based on the belief that it would have to pay the rates and benefits defined in the more recent Order. The other contractors and subcontractors bidding on the project shared this belief. So did the owner.

After the contract was signed, the contractor discovered that the contract documents, prepared by the owner's architect, obliged it to follow the Act, not the Order. Vic Van Isle sought legal advice on how to deal with the error and was told that it was bound by the terms of the bid — as it was bound when it made the error in its previous bid — and was, therefore, required to follow the Act. In the words of the trial judge:

> The [contractor] proceeded accordingly, presumably on the reasonable supposition that the law in Kelowna was the same as the law in Williams Lake.

The architect, however, ordered the contractor to comply with the Order. Vic Van Isle asked the architect for a change order to cover the extra costs. The architect refused. The contractor completed the project under protest and sued the owner for the difference between the fair wages it paid pursuant to the Order and what it would have paid pursuant to the Act.

At trial, Justice Wilson followed the reasoning in *Ron Engineering*: The owner's invitation to contractors to submit bids was an offer. Part of the offer was the requirement to pay fair wages pursuant to the 1976 Act. When the contractor accepted the owner's offer and submitted its bid, Contract A came into being. Compliance with the Act was one of the terms of that contract.

When the parties entered into the construction contract, or Contract B, they thought that wages and benefits were to be paid pursuant to the Order. But no matter what they believed, they were bound by the express terms of Contract A.

The judge quoted from the decision of the Alberta Court of Appeal in *Calgary v. Northern Construction Co.*:

> The entering into Contract B would be, in actuality, a mere formality, for all of its terms were provided for in Contract A expressly or by reference to other documents. Unless a contractor knew all of the terms that were to be incorporated into Contract B, he would be unable to determine a bid price.

For the same reason, decided the judge, the fair wage provisions of Contract A were an integral part of Contract B. Thus, the contractor was right. Contract A stipulated compliance with the Act, and when the architect required it to follow the Order, that was a change in the work under the contract.

Still, the contractor did not win any damages at trial. The court found that the rates and benefits stipulated by the Order were an implied term of Contract A. The Act required that the contractor pay wages and benefits usually provided to workers on public construction projects in the geographic area.

The wages usually paid in the area, decided the judge, were as defined in the Order. Therefore, when the architect ordered the contractor to pay such wages, the change did not entail any extra costs.

The Court of Appeal had a lot of sympathy for the contractor:

> One would have thought that, on the face of it, the [contractor] would win hands down. It had accepted the contract freely proffered by the School District. It had done nothing to induce the School District to act to its detriment in the tendering process. It had not had any part in the wording of the contract. It had not breached the contract or offended the mechanics of the tendering process. Indeed, the bargain the parties struck was, in commercial terms, entirely satisfactory to each.

The difficulties which arose did so not from the contract but from external factors which had nothing to do with performance in accordance with the specifications and at the agreed fixed price.

Furthermore, the appeal court rejected the term implied by the trial judge. In principle, courts must not imply a term unless it can be shown that such a term is necessary to give the contract business efficacy. The Court of Appeal did not find the "necessity factor" in the case before it.

For 16 years prior to 1993 contracts had been successfully performed "in lock step with the Act". There was nothing unique in the circumstances of the Vic Van Isle contract that made it necessary to imply a term to enable it to do likewise.

Justice Gibbs and Justice Hall both quoted from the 122-year-old decision of the House of Lords in *Inglis v. Buttery*. There, Lord Blackburn said:

> ... where parties agree to embody, and do actually embody, their contract in a formal written deed, then in determining what the contract really was and really meant, a Court must look to the formal deed and to that deed alone. This is only carrying out the will of the parties.

When Vic Van Isle discovered that the contract allowed it to pay the lower rates, that represented an unanticipated prospective financial gain for the contractor. In strict commercial terms, that did not make any difference to the School District. The price for the school project remained unchanged. However, when the architects required compliance with the Order, in spite of the contract, the effect was to divert funds that would have been the property of Vic Van Isle to the construction workers.

"What occurred was entirely due to errors made by or on behalf of the School Board during the tendering process", said Justice Gibbs. He concluded: "I see no injustice in requiring the School Board to live with its own errors."

However, Justice Gibbs constituted a minority of one. The majority of the Court of Appeal, Justices Hall and Rowles, preferred the result reached by the trial judge, even though they rejected his reasoning. They proposed to reach that same result, but by a different route.

After tenders closed, Vic Van Isle and the owner entered into negotiations to reduce the price in accordance with the owner's budget. When they finally reached agreement, they both understood that the *Fair Wage Order* was applicable to the job. That was a new agreement, and that was what the majority of the Court of Appeal decided to enforce. In this they relied on the equitable principle of *estoppel*.

---

**Estoppel**. The doctrine applies when one party to a contract leads the other (by words or conduct) to believe that it will not enforce some or all of its rights under the contract. If the other party relies on this belief and alters its activities accordingly, the principle of estoppel may prevent the first party from enforcing the original contract in court.

---

The court dismissed the contractor's appeal. The contractor then sought leave to appeal to the Supreme Court of Canada, but was refused.

The court concluded with the statement, repeated in one form or another in every tendering decision since *Ron Engineering*, that "it is vital that the integrity of the bidding process be preserved".

---

*It appears that the courts will not always strictly enforce the terms of construction contracts in the bidding context. There is still some room to manoeuvre. The doctrine of estoppel will allow the courts to go around contractual terms when they come to the conclusion that those terms do not represent the real understanding between the parties.*

*The following important case does not involve a mistake, but a difference of opinion. It offers guidance in the difficult situation when the bidder and owner have a serious disagreement regarding some term of the bid documents.*

# NORTH VANCOUVER (DISTRICT) v. PROGRESSIVE CONTRACTING (LANGLEY) LTD.

Court of Appeal for British Columbia; April 1993

A contractor did not repudiate its bidding contract when it had a legitimate and honestly held difference of opinion with the owner over the meaning of a term in the contract, decided Justice Hall of the Supreme Court of British Columbia. The Court of Appeal agreed.

In January 1990, the District of North Vancouver issued an invitation to contractors to bid for the construction of a road. Each bid had to be accompanied by a certified cheque or bid bond equal to 10% of the bid. The lowest bidder was Progressive Contracting (Langley) Ltd.

The largest item in the lump-sum price was "road construction to 20 mm crushed gravel standard". Progressive's bid under this heading was $65,000 while the next lowest bid was over $130,000. Almost immediately after the opening of the bids, it became apparent that there was a fundamental misunderstanding between the owner and Progressive regarding the cost of the supply of gravel under the contract.

The owner planned to provide gravel to the contractor through its own pits at fixed prices and planned to back-charge the cost later. Progressive, on its interpretation of the contract documents, did not expect a deduction for the cost of the gravel from its lump-sum bid.

When it discovered the disagreement, Progressive requested that it be allowed to withdraw its bid without forfeiting the bid bond. When the owner refused, the contractor asked to have the prices for gravel renegotiated. The District refused to renegotiate. Instead, it formally accepted the bid as submitted and sent the construction contract to Progressive for signing.

It warned the contractor that if it did not execute the contract, the owner would consider this a fundamental breach of the contract. The contractor signed the contract and sent it back with a letter. The letter expressed its disappointment that the owner was "not prepared to resolve amicably the ambiguity that appears in the tender documents and is brought forward in the formal contract documents".

The letter also advised the owner that the contractor's interpretation of the agreement had not changed. The owner replied:

> Please be advised that your letter is being treated by the District as notice that you intend to not abide by the terms of the contract awarded to you and that therefore that contract is repudiated. The District wishes to formally advise you that your repudiation is accepted by it and a call will be made on your bid bond. The referenced contract will be awarded to the next low bidder. If there is any shortfall in

the security provided under your bid bond and the costs of the work under a new contract ... those costs will be recovered from you as well.

At trial, the owner argued that Progressive's bid was duly accepted, and, therefore, the contractor was bound to enter into a contract to do the work. Although the contractor signed the contract and returned it with its letter, the letter amounted to a *repudiation* of its contractual obligations. Further, the contractor's position was so fundamentally at odds with its obligations to perform under the contract that, in fact, it did not truly comply with its terms.

> **Repudiation of Contract.** An indication by a party to a contract that it will not perform its obligations in the future and so commit a breach of contract. Repudiation may be expressed in words or implied from conduct. The repudiation of a contract entitles the other party to treat the contract as at an end and to sue for damages. Treating the contract as at an end in response to repudiation is sometimes also called repudiation, as shown in the quote below.

Progressive maintained that although it disagreed with the owner as to the interpretation of the terms of payment for gravel, it was prepared to execute the construction contract, to undertake the work and to try to persuade the owner in due course to accept Progressive's interpretation.

Justice Hall found that some disagreement as to the meaning of a term or terms in a contract may well not amount to repudiation. The leading authority on the subject is the 1874 decision of the House of Lords in *Freeth v. Burr*. The top court in England said (emphasis added):

> It is not a mere refusal or omission of one of the contracting parties to do something which he ought to do that will justify the other in repudiating the contract; but there must be an *absolute refusal* to perform his part of the contract.

In reviewing other cases, Justice Hall said that the courts must be concerned to ensure that parties live up to their contractual obligations. A party cannot pretend to be willing to perform but, in fact, propose to take action to the contrary. If, for instance, Progressive had said, "we will execute the contract but only perform it a year from now", that would amount to repudiation.

A party will not be allowed to adopt a position that would wholly frustrate the performance of its obligations under a contract, but a legitimate and honestly held difference of opinion concerning a term of a contract is not necessarily repudiation. Said the judge:

It does not appear to me that in this case Progressive was refusing to enter into the contract or refusing to undertake its obligations under the contract. It signed the contract document and indicated its readiness to proceed. It alerted the [owner] to the fact that it was taking a certain view of matters as to the basis for compensation for gravel works. The [owner] apparently felt that it was not prepared to accept this position of Progressive and elected to treat the stance of Progressive as repudiation.

Counsel for the owner tried to convince the court that the contractor was only showing a token agreement to proceed with the construction contract. Justice Hall perceived the situation differently. He was convinced that the contractor was acting in good faith even though it faced a difficult argument in view of the wording of the contract.

He found it significant that the contractor never refused to enter into the contract. Therefore, he held that Progressive did not repudiate the contract — the owner repudiated it by failing to sign the construction contract and failing to allow Progressive to do the work. He dismissed the owner's action and awarded damages to the contractor.

The owner appealed. Justice Goldie, writing for the Court of Appeal, agreed with the trial judge that the contractor had not repudiated the contract. When the owner informed Progressive that it had accepted its bid, he said, the contractor had two choices:

- It could refuse to enter into the construction contract on the basis that its view of the bid documents was right and that of the District was wrong; if the District subsequently proved to be right, the contractor would be in breach of Contract A.
- It could enter into the contract, proceed with the work and seek whatever remedy was open to it under the contract.

Progressive chose the latter approach. It executed Contract B and took on the work. It continued to stress that its version of the bid documents was right, but, said the judge, this was irrelevant to its contractual obligations under Contract B.

The District apparently believed that Progressive could not maintain its rights under Contract A while agreeing to perform Contract B. "In my view", said Justice Goldie, "the one could have no bearing on the other".

In general, if two parties sign a contract and then it turns out that they disagree on the terms, the contract still stands. As Lord Denning said in *London County Council v. Henry Boot and Sons*:

> It does not matter what the parties, in their inmost states of mind, thought the terms meant. They may each have meant different things. But still the contract is binding according to its terms — as interpreted by the court.

In the opinion of Justice Goldie, the case before him was even clearer:

There was no disagreement over the terms of Contract B. Progressive was ready to perform that contract. The District repudiated it.

The other two judges hearing the appeal agreed and dismissed the appeal.

---

*The following case is an example of the severe consequences of a latent (invisible) mistake in a bid. The appeal court decision was too late to save the bidder from destruction following a harsh decision at trial level.*

## CITY OF OTTAWA NON-PROFIT HOUSING CORP. v. CANVAR CONSTRUCTION (1991) INC.

Ontario Court of Appeal; April 2000

In 1991, Canvar Construction (1991) Inc. submitted a lump-sum bid to the Ottawa Non-Profit Housing Corporation (ONPH) for the construction of a residential building. Together with the bid, it submitted a bid bond of 5% of the bid price, as required by the owner. The bid was irrevocable for a period of 60 days after the date of the bid closing.

Canvar intended to bid $2,989,000 and so its bid bond was in the amount of $149,450, that is, 5% of the intended bid amount. However, due to a clerical error, the bid itself as written on the bid form was only $2,289,000 — $700,000 less than it should have been.

Even at the intended price Canvar was considerably lower than the next lowest bidder whose price was $3,130,000. The owner's estimate of the cost of the project was $2,887,535.

At the opening of the bids Canvar realized it had made a mistake in the bid price and immediately pointed it out to ONPH. The contractor asked to be allowed to adjust the price or to withdraw its bid.

ONPH rejected both options, and so Canvar refused to execute the construction contract. ONPH then awarded the contract to the next lowest bidder and sued Canvar for the difference of $841,000. In the alternative, it claimed damages of $726,650 plus the amount of the bid bond.

Justice Chilcott distinguished Canvar's case from the precedent decision of the Ontario Court of Appeal in *McMaster University v. Wilchar Construction Ltd.* In that case, the document submitted was missing one page and was, therefore, evidently incomplete. Nor was the situation similar to another case where an offeror intended to say $200 a ton but wrote $20 by mistake.

The judge found no evidence that Canvar did *not* intend to submit the bid in the form and substance it was in. The fact that the bid bond was more than 5% of the bid price was no indication of what was intended, since the contractor was free to submit a higher bid bond than the minimum required. In fact, one of the other bidder deposited a bid bond of 10%.

The Contract A as defined in *Ron Engineering* was formed when ONPH asked for bids and Canvar accepted this offer by submitting its bid. Using a line of reasoning that contractors find very hard to accept, the judge held that there was no mistake in the formation of Contract A, because Canvar intended to submit the bid *in the form* in which it was submitted.

Justice Chilcott relied on the milestone case *Calgary (City) v. Northern Construction Co. Ltd.* where the facts were very similar to those of the *Canvar* case. The court in that case refused to allow the contractor to correct its bid, even though the corrected bid would still have been lower than any other, and said:

> It is not unconscionable for a tenderee [owner] with knowledge of motivation error [in the bid] to hold a tenderer to his bargain provided that his doing so does not impose a grossly disproportionate burden upon the tenderer.

The owner had a duty to mitigate its damages, added the judge, and it fulfilled that obligation by accepting the second lowest bid.

With regard to damages, the bid form gave the owner the right to retain the contractor's deposit and to accept the next lowest or any bid. The bid further stated:

> We also agree to pay the corporation [ONPH] the difference between this tender and any greater sum which the said corporation may expend or incur by reason of such default or failure....

Clearly, said the judge, ONHP was entitled to the amount of the bid bond, as well as the difference in bid price, namely $841,000.

Canvar argued in its defence that the owner's claim was unconscionable. The ingredients to establish ro prove unconscionability were set out by the Court of Appeal in *Harry v. Kreutziger*:

> Where a claim is made that a bargain is unconscionable, it must be shown for success that there was inequality in the position of the parties due to the ignorance, need or distress of the weaker, which would leave him in the power of the stronger, coupled with proof of substantial unfairness in the bargain. When this has been shown a presumption of fraud is raised and the stronger must show, in order to preserve his bargain, that it was fair and reasonable.

Justice Chilcott found that such ingredients were not present in the *Canvar* case. But what of the court's reservation in the *Calgary* case regarding imposing "a grossly disproportionate burden upon the tenderer"? In that case, the difference between the bids was about 4%; the difference between the Canvar bid and the next lowest was 36.74%. Still, no authority could be found as to the meaning of the words "disproportionate burden" other than speculations by both counsel. The judge therefore found that damages of $841,000 was not a disproportionate burden on Canvar.

The *Calgary* case was affirmed in the Supreme Court of Canada, and was followed in Ontario in the case *Vaughan (Town) v. Alta Surety Company*. Justice Chilcott found himself bound by the reasoning in the *Calgary* case, and accepted the reasoning in *Vaughan*.

The court therefore assessed the damages at $841,000. Of this, the contractor's surety company which issued the bid bond had to pay $149,450.

The reasons for judgment at trial do not include an extensive review of the evidence presented. However, on the face of it, *Canvar* looked distinguishable from *Ron Engineering*.

In *Ron Engineering*, the contractor's bid was close to the consultant's pre-bid estimate. In *Canvar*, the contractor's bid was 36% under the second lowest bid and some 25% below the owner's pre-bid estimate (and fondest hope). One obvious distinction from *Ron Engineering* is the relationship of the bid to the pre-bid estimate.

The trial decision directed one sentence to the discrepancy between the Canvar bid and the pre-bid estimate. The trial judge said:

> The estimate of the work done by the plaintiff's consultant would not indicate a gross miscalculation in the tender price.

For sure, evidence was led on this point. But, the reasons provide little insight as to why this significant price discrepancy carried no weight. The notion of fairness did not arise in the reasons of the trial judge although he did consider and then reject the idea "that the transaction in this case was unconscionable".

The sum of $841,000 left on the table in a $30,000,000 job would not be that remarkable. In Canvar's $3,000,000 dollar job, it was a disaster. Enough of a disaster to put Canvar out of business. So, when it came to launching an appeal, Canvar's bonding company – with $149,000 in play – was the real appellant.

The Court of Appeal read the entrails differently. The bid documents called for bid security equal to 5% of the amount bid. Canvar submitted bid security in the amount of $149,450 – 5% of what it claimed it intended to bid. The Court of Appeal connected the dots and determined that it should have been obvious to the owner that Canvar's bid was in error. No contractor, it reasons, submits bid security in excess of what is required. They just don't behave that way.

Having come to this conclusion, the Court of Appeal was drawn to the *Wilchar* case which the trial judge had distinguished. The Court of Appeal ruled that Canvar's error was apparent on the face of its bid and, relying on the clear exception made in *Ron Engineering*, ruled that the owner was not entitled to accept the bid it knew (or should have known) was mistaken. Canvar's bonding company was off the hook. Of course, Canvar itself was still "a goner".

Since Ron Engineering, the Contract A/Contract B model has been in the Supreme Court of Canada four times, three of them since 1999. The Court has reconsidered the Ron Engineering model but it has not encountered a latent mistake case since the 1980s.

Is there anything in the Supreme Court of Canada decisions or in the pronouncements of lower courts, to suggest that the outcome of Ron Engineering might be different if the same case were tried today? The answer to that question is no. The most recent illustration lies in the Canvar case.

Was the court decision fair? Good question. Was it predictable? As to the law, yes it was.

The implied duty of fairness which now exists in all bid documents, unless expressly excluded, has not been brought to bear in the case of latent error — like Ron Engineering. If the argument were advanced, it would probably fail.

Is it fair for a bidder to suffer a huge loss because of a clerical error? In global terms, it probably isn't fair. But, within the narrow confines of Contract A, the mistaken bidder has accepted an offer — and entered a contract — where one of the terms and part of the consideration is that the bid is irrevocable for a defined period of time. So, in accepting the offer for Contract A, the bidder knows — and accepts the risk — that a latent error in its bid is something it will have to live with, fair or not.

For those wondering whether Ron Engineering — to the extent that it dealt with latent mistake — was alive and well, rest easy. The same case law was applied at trial and on appeal. But, the outcome changed because those two courts each had a different fix on the facts.

Having heard the evidence, the trial judge decided that there was no error "apparent on the face of the bid". The appellate court would not have interfered with this ruling had it not believed that the trial court missed the mark by a wide margin. But, there was no difference in the two courts as to the application of Ron Engineering.

The reason that there are few reported cases on latent mistake is that the law is clear. The risk of an error that is not "apparent on the face of the bid" lies with the contractor — still.

---

Twenty years after Ron Engineering, the construction industry still had not exhausted the supply of fresh ideas in adapting to the new rules of the bidding game. In the case of Derby Holdings Ltd. v. Wright Construction Western Inc., the bidder tried to save itself from the consequences of a major mistake in its bid by claiming the bid was non-compliant. The owner attempted to snap up the bid nevertheless, on the basis that part of its own bid call was invalid.

## DERBY HOLDINGS LTD. v. WRIGHT CONSTRUCTION WESTERN INC.

Saskatchewan Court of Appeal; September 2003

Derby Holdings Ltd. invited bids for renovations to a shopping mall in Saskatoon. Four general contractors responded before the deadline on May 28, 1999. Wright Construction Western Inc. submitted a bid of $1,347,130. The next lowest bid by Gabriel Construction Ltd. was $1,656,360; the other two bids were $1,735,246 and $1,793,081 respectively.

When Wright reviewed its bid, it determined that it had made a significant error in compiling the electrical portion. The mistake amounted to $208,939. In a letter dated May 30, Wright informed the owner's architect, Ken Wilson, of the mistake, and offered to negotiate a price adjustment that would increase the amount of its bid yet keep it below the amount of the next lowest bid.

Derby refused to negotiate, and purported to accept Wright's bid. Wright refused. In the words of the judge at the subsequent trial:

> In short, [Wright] was not prepared to honour its bid because of its own mistake, while [Derby] was not prepared to give up the windfall it would receive because of that mistake.

Eventually, Derby accepted the next lowest bid and the project was completed. The lawsuit followed. Derby claimed $309,230 – the difference between Wright's bid and that of the next lowest bid.

In order to reach his decision, Justice Baynton of the Saskatoon Court of Queen's Bench had to unravel a convoluted bidding situation.

The first complication was a Memo from an employee of Ken Wilson, Derby's architect, faxed to each of the prospective bidders on May 26. It contained the following paragraph:

> Note: the Electrical portion of the work as indicated in the specifications and on the drawings HAS BEEN TAKEN OUT OF THIS TENDER, AND WILL BE TENDERED SEPARATELY in a few days, once appropriate changes have been made.

This Memo was followed by Addendum #1, issued by Ken Wilson himself later the same day. The Addendum contained seven pages of specification revisions and 14 pages of drawings. It told the bidders that they must disregard the notice sent out in the morning, and that the Addendum must be incorporated into the contract documents. It required the bidders to acknowledge receipt of the Addendum on the bid form. It stated very emphatically that the electrical work *was* part of the bid.

Clause 18 of the Instructions to Bidders issued as part of the bid documents read:

> Addenda may be issued during the tendering period, *but not later than 3 days before the tender closing*. Addenda issued become part of the Tender Documents, and instructions therein will be included in the Construction Contract. *Each bidder must acknowledge the recent of such addenda on the Tender Form. [emphasis added]*

Both the Memo and Addendum #1 were issued on May 26, less than three days before the May 28 bid closing. The obligatory bid form contained no space or line for the acknowledgement of any addenda.

Neither Wright nor the next lowest bidder acknowledged receipt of the Addendum in their bids. After opening the bids, Wilson telephoned Wright to find out whether Wright had received the Addendum. Wright confirmed that it had; nothing else was said at the time.

The evidence at trial demonstrated the lack of communication between Derby and its architect over many issues. In the words of Justice Baynton, "To put it mildly, none of the participants in the project were on the same wave length...". The confusion surrounding the bidding process was pervasive:

- Two bidders failed to acknowledge Addendum #1, one "acknowledged" it and the remaining one "acknowledged receipt" of it.
- One term of the instructions to bidders provided that *all* addenda became part of the project to be bid while another term provided that addenda could not be issued after three days prior to bid closing.
- The architect considered that because a term of Addendum #1 instructed prospective bidders to disregard the previous Memo, that Memo had no further legal effect. But both amending documents were issued after the time had expired for issuing them. This in turn raised the significant legal issue as to whether the two documents had any binding effect whatsoever.
- At one point it was suggested by Wright that because the Memo was not an addendum, it was effective while the Addendum #1 purporting to cancel it was not, and that accordingly the whole of the electrical component was not included in the bids.
- When Ken Wilson telephoned Wright for clarification, he did not ask whether Wright's bid *included* the work set out in Addendum #1 but only whether Wright had *received* the Addendum, and that was what Wright confirmed. The architect accordingly could not be sure, even after making the phone call, whether Wright's bid included the work set out in Addendum #1.
- No call was made to Gabriel Construction, the second lowest bidder, to clarify this issue. The architect simply assumed that Wright's bid included Addendum #1, and that it accordingly had to be the lowest bid.

In court, Wright's principal defence was that it had no liability to Derby. Wright maintained that its bid could not have been accepted because it was non-compliant – it did not include the work comprised in Addendum #1, and did not specifically acknowledge that Addendum on the bid form as required by the instructions to bidders.

Derby took the position that neither the Memo nor Addendum #1 had any legal effect because they were issued after the deadline. Furthermore, the telephone acknowledgement of Wright was a rectification of the non-compliant aspect of Wright's bid. Wright's bid was irregular but not informal, and therefore was capable of acceptance.

Derby also had an alternative position. If Wright's bid was non-compliant, argued Derby, then it constituted a counter-offer that was open for acceptance by Derby. Once the counter-offer had been accepted by Derby, there was a contract and Wright was liable for its breach.

Wright's argument caused Justice Baynton some headaches. At the conclusion of the trial, he favoured the position of Derby, primarily because he suspected that the non-compliance was a "smokescreen" raised by Wright to obscure its real reason for refusing to honour its bid, namely the mistake. In the end, he reluctantly changed his view of this issue. He decided that Wright's bid was non-compliant, and therefore could not be accepted.

The failure of Wright to acknowledge the Addendum in its bid was not merely an irregularity. A crafty bidder might decline to acknowledge an addendum, said the judge, in order to get a chance to fine-tune its bid once the other bids were known. It could then elect whether to verbally acknowledge or refute the addendum depending on the economic consequences. If the law condoned that practice, it would seriously jeopardize the integrity of the bidding process.

Nor could an otherwise non-compliant bid be made compliant by addressing the deficiencies after the bids are opened, as decided in the case *Vachon Construction Ltd. v. Cariboo (Regional District)*.

Finding that a bid was non-compliant at the behest of the contractor whose bid was accepted by the owner was new law, acknowledged the judge. But, he added, there is no valid reason why the application of this legal principle should depend on the source of the challenge. Nevertheless, Judge Baynton was not quite at peace with his own decision:

> I am troubled that my decision on this issue appears to have the potential of permitting a contractor to avoid its promises and obligations by its own neglect or by its failure to submit a tender that substantially complies with the invitation to tender package.

The judge made short shrift of Derby's alternative claim that Wright's bid was a counter-offer and that this counter-offer was accepted by Derby. Such a proposition flies in the face of the special principles of contract law that apply to the bidding process that have been developed since *Ron Engineering*, he said. The implied term of Contract A Derby had with *all* the bidders precluded it from accepting a non-compliant bid under any guise.

Wright's error was honestly but negligently made. On the basis of the *Ron Engineering* case, Wright would nevertheless remain liable to Derby on Contract A (had the bid been compliant) but would not be obligated to enter into Contract B (the construction contract). Even an honest mistake will not relieve the bidder of its obligations under Contract A, but it will relieve the bidder of its obligations to enter into Contract B.

In any case, even if Wright's bid had been a counter-offer capable of acceptance, Derby knew that a significant mistake had been made in the price. The law of mistake would preclude a binding contract.

The court dismissed Derby's action, but Wright failed to win its costs. Had Wright not submitted a defective bid, there would have been no lawsuit. Its actions afterwards only exacerbated the problem. "In the circumstances, it would be inappropriate to reward [Wright] with an order of costs," concluded Justice Baynton, and ordered that each party bear its own costs.

The Saskatchewan Court of Appeal agreed with Justice Baynton's decision. It added a brief comment regarding Derby's argument that the Memo and the Addendum were both invalid as they were issued less than three days before the closing date for bids, something not permitted by the bid documents. That position, said the court, excludes the possibility of taking into account evidence of custom in the industry permitting late changes, of which there was some evidence at trial. It was therefore open to the trial judge to find, as he did, that the bidding process was so flawed by uncertainty as to what was to be included in the bids, that it could not give rise to a contractual relationship, and that Wright's bid could not be a compliant bid.

---

*The Derby case is a forerunner of many cases where a contractor escapes Contract B by arguing that Contract A did not come into being because its bid was non-compliant. And why not?*

*Does it matter which party discovers that the other has failed to communicate an acceptance of the offer for Contract A? It shouldn't. The offer for Contract A has either been accepted or it hasn't. The real question is whether the three ingredients — offer/acceptance/consideration — exist permitting a contract to arise.*

*In Derby, and cases like it, the non-compliant bid may still be a valid offer – for something other than Contract B. An offer like the one made by the contractor in Derby is an old fashioned "nudum pactum" (an irrevocable offer,* with no consideration) which, even though it is irrevocable, can be withdrawn before acceptance on the grounds of mistake.

The Derby case stimulated owners to modify bid documents so as to close the non-compliance loophole for mistaken bids. As a result, many such documents contain — in addition to the typical privilege clause reserving the right to accept or reject any bid — a new kind of disclaimer: a **"discretion clause."** This one, in so many words, purports to reserve for the owner the right to accept a non-compliant bid.

The following case is the first where such a clause was tested in an appellate court, and found wanting.

## GRAHAM INDUSTRIAL SERVICES LTD. v. GREATER VANCOUVER WATER DISTRICT

British Columbia Court of Appeal; January 2004

Graham Industrial Services Ltd. was the successful bidder on a project for the Greater Vancouver Water District. Its $21,500,000 bid was the lowest of four by about $5,000,000.

After bids were opened, Graham advised the District that the bid contained a $2,000,000 error, and requested that it be allowed to withdraw. It also claimed that the bid was non-compliant, and therefore incapable of acceptance in any event.

The District nevertheless accepted Graham's bid. Graham went to court and asked for a determination of the issue. It produced a long list of "non-compliant" items in its bid.

The District, relying in part on advice from independent consultants, determined that none of these alleged shortcomings of Graham's bid were material, and relied on a "discretion clause" in the Instructions to Tenderers:

> If a Tender contains a defect or fails in some way to comply with the requirements of the Tender Documents, which in the sole discretion of the Corporation is not material, the Corporation may waive the defect and accept the Tender.

To make the owner's intentions quite clear, the Instructions to Tenderers also contained an elaborate "privilege clause" in essence reserving the right to accept or reject any bid.

The chambers judge, Justice Edwards, quoted from the Saskatchewan Queen's Bench decision (upheld by the Saskatchewan Court of Appeal) in *Derby Holdings Ltd. v. Wright Construction Western Inc.*, where the court found that "there is no justification for a rule of law that permits an owner to hold a tenderer to a bid that the owner itself has pre-determined to be non-compliant."

CHAPTER 6: MISTAKE IN BID **143**

The chambers judge, Justice Edwards, gave most of the alleged non-compliance items in Graham's bid short shrift. He found them insignificant. He even found that hand-printed names of two officers of the company constituted signatures, contrary to Graham's argument that the document was unsigned.

"It can be confidently inferred," concluded the judge, "that but for its own error Graham, had its winning tender been challenged by a competitor, would have argued that the alleged shortcomings in its own tender it now asserts are fatal, could be overcome by the District relying on the privilege and discretion clauses."

However, two sections of Graham's bid which called for technical submissions were in a different category from the rest. Graham's tendered response to Section 8.6 was about half a page in length and included the statement "We will comply with the specifications as prescribed ..." without elaboration.

Graham's response to Section 8.7 requiring an Environmental Protection Plan was very short, and ended with the following statement: "We will prepare a project specific plan ... following award of the contract."

The court found that Graham's responses to Sections 8.6 and 8.7 of the Tender Form were so patently deficient they could not, on an objective reading, be said to "conform in all material respects to the Invitation to Tender."

Justice Edwards concluded:

> The District's authority under the privilege and discretion clauses to determine non-compliance with Tender Form requirements ... must be exercised both in good faith and in manner which can withstand *objective scrutiny*. Otherwise those clauses could be used to deem a non-compliant tender to be a compliant tender. That would undermine the fairness of the tendering process. Preservation of that fairness is the underlying rationale for the requirement for substantial or material compliance. *[emphasis added]*

In other words, compliance cannot be conclusively determined *subjectively* by the owner, it must also pass the objective scrutiny of the court. In view of Graham's material non-compliance with Sections 8.6 and 8.7, the District could not accept Graham's bid, never mind the discretion and privilege clauses. No Contract A was concluded, and Graham was off the hook.

On appeal, the District contended that the chambers judge erred when he substituted his own analysis as to the adequacy of Graham's bid for that of the owner, contrary to the express terms and conditions in the instructions to bidders. In my opinion, stated Chief Justice Finch writing for the Court of Appeal, the chambers judge was correct to apply an objective test to the question of compliancy." He refused to interfere with his application of that test and his conclusion that, according to an objective analysis, Graham's bid was materially non-compliant. He said:

> In my view, the Discretion Clause in the instructions to tenderers cannot permit the Water District to determine subjectively whether a defect is material.

The District's right to rely on the discretion clause as a term of Contract A only arises if a valid Contract A is formed, said the Chief Justice. Contract A is only formed if a bid is compliant. Since the discretion clause does not operate before Contract A is formed, the determination of whether a bid is capable of acceptance in law must be based on an objective analysis. A not-yet-operative discretion clause cannot give the owner the power to decide that a bid is compliant if, "on an objective analysis", the bid is materially non-compliant.

In *Kinetic Construction Ltd. v. Comox-Strathcona (Regional District)*, Justice Preston of the B.C. Supreme Court found that there is no principle of law requiring the owner to reject a non-compliant bid if the bid documents expressly reserve to the owner the right to accept such bids.

Chief Justice Finch did not feel he had to consider whether Justice Preston was right or not. In *Kinetic*, the non-compliant bidder withdrew its "counter-offer" *after* the owner had accepted it. The discretion clause in that case had therefore become operative. Graham, on the other hand, revoked its counter-offer *before* the District purported to accept its bid so no legal contract could have possibly resulted.

In any case, it appears that the judge had misgivings about the discretion clause quite apart from the issue of its being operative or not:

> In my view, giving the Discretion Clause the effect for which the Water District contends would allow the Water District and other owners to circumscribe the tendering process. The mandatory requirements of the instructions to the tenderers would be completely negated if the Water District had the right to exercise its discretion to waive any defect or non-compliance by deeming material omissions to be non-material.

What meaning, then, is to be given to the discretion clause? Chief Justice Finch explained its function:

> In my view, the clause simply recognizes that the test for determining whether a tender is valid is one of *substantial compliance* rather than strict compliance. The clause allows the Water District to accept tenders with minor irregularities or non-material defects. *[emphasis added]*

This substantial compliance test, he said, is consistent with an objective analysis of whether Contract A has arisen. The unanimous Court of Appeal rejected the District's appeal.

---

*The main messages from the B.C. Court of Appeal in Graham were that substantial compliance is the threshold to entering Contract A, and, the determination of whether a bid defect is material enough to render it non-compliant is open to objective review. Right on!*

*The Court of Appeal concluded that the "discretion clause" in the bid documents "does not operate" because Graham did not have Contract A with the owner. The no "Contract A" part was true, given that the Court of Appeal held that Graham's bid could not be said to be substantially compliant.*

*The "discretion clause" needs to be viewed from two vantage points. From Graham's perspective, the clause is a term of the owner's offer to enter Contract A. The owner is the gatekeeper and its offer of Contract A included a discretion to determine whether a bid defect is or is not "material".*

*From the perspective of the bidders who do have Contract A with the owner, the "discretion clause' is a term of Contract A. Those other bidders are entitled to call upon the owner to account for the manner in which it exercised its discretion. After all, those bidders have an interest in seeing that a non-compliant bidder — one which may have a significant advantage over them due to its non-compliance — is not admitted to Contract A.*

*If the last two points are so, what effect is given to the "discretion clause" which gives the owner "sole discretion"? Here again, we need to look at the clause from within Contract A and from outside it.*

*For those who do have Contract A with the owner, the "sole discretion" means that the owner does not have to consult with the bidders who have Contract A in determining what is and isn't "material". If the determination by the owner of "material" was close to what the industry might accept, the owner could claim the benefit of the doubt — and get it. But, the clause does not say that the word "material" has any meaning that the owner wants to put on it.*

*From the perspective of Graham, the "sole discretion" also suggests that the owner will make the decision without consultation. If the discretion were interpreted any more broadly, then the owner — as it tried to do with Graham — could decide to forego a defect which the whole world thought was "material" but which the owner, for its own selfish reasons, decided to treat as "immaterial". Allowing the owner to cancel the law of mistake — you cannot create a contract by snapping up an offer you know to be mistaken — was not something the Court of Appeal was prepared to do. What isn't entirely clear is why the Court needed to decide that the "discretion clause" did not operate from Graham's end of the equation. It doesn't seem right, does it?*

---

*The thrust-and-parry over mistaken bids was not exhausted with the non-compliance/discretion clause encounter. The following case is the first that used another novel approach to Contract A. In the cases seen so far, the mistaken bidders typically argued that (a) the bid contained an error on the face of the* bid, or (b) that the bid was non-compliant; either way, they disputed the existence of Contract A.

The new approach accepts that Contract A may have come into existence when the mistaken bid was submitted but asks the court to apply a very old equitable remedy for contracts that contain a unilateral mistake i.e. one where only one party is mistaken.

## TORONTO TRANSIT COMMISSION v. GOTTARDO CONSTRUCTION LIMITED

Ontario Superior Court of Justice; December 2003

In October 2000, the Toronto Transit Commission (TTC) issued a call for bids for the construction of a bus garage. Gottardo Construction Limited submitted a bid of $4,811,000 but, within one hour after its bid was opened, the contractor advised the TTC that it had made a mistake of $557,000 — before the TTC had done anything with the bids.

Gottardo's letter to the TTC concluded with the following:

> We hereby request that we be permitted to withdraw the Tender without penalty. Please take no action against our bid bond. To do otherwise would cause severe and irreparable financial hardship to us. This was an honest and inadvertent error.

The TTC notified Gottardo, as the bid documents provided, that Gottardo deliver certain project details described as Tender Submission Information. Gottardo did not make the required submission.

The TTC found nothing on the face of the original bid to indicate that an error had been made, and therefore, as a matter of policy, it informed Gottardo that its bid had been accepted. When Gottardo failed to enter into a construction contract, the TTC awarded the contract to the second lowest bidder, at an increase of $434,000 over Gottardo's bid.

In April 2001, the TTC sued Gottardo for the penalty amount of the bid bond of $520,000 and, in the alternative, for damages of $434,000, namely the differential between the two low bids. Eventually, the TTC abandoned its claim on the bid bond and pursued only the damages claim.

Justice Kiteley of the Ontario Superior Court of Justice found that Gottardo had indeed made an honest and inadvertent mistake, and that the fact was communicated to the TTC before it had "altered its position in reliance on the tender."

Gottardo argued that Contract A had not come into existence (and therefore it was not obliged to proceed with Contract B) because:

- there was an error on the face of the bid;
- the bid was otherwise non-compliant;
- if Contract A was created, it should be subject to rescission.

To consider the first two grounds of Gottardo's defence, it is important to understand the bid process which the TTC created. First, the process defined a family of documents which constituted the "Tender Documents". Within that definition was a subset of documents called "Tender Submission Information". At closing, the bidder was to submit its price and associated materials such as its bid bond and agreement to bond. Then, within two days of a written request from the TTC, a bidder was obliged to submit the Tender Submission Information.

The trial judge held that the manner in which the TTC had defined Tender Documents had, in effect, created a two-stage closing. The first stage was for all bidders. The second stage was for those who were required to submit the Tender Submission Information.

The trial judge found that Gottardo was compliant on the first submission. If this had been a one-submission tender, that would have been that. *Ron Engineering* all over again.

But, having created a two-stage process, the TTC was stuck with it. Having discovered its mistake in pricing, Gottardo (wisely) failed to provide the Tender Submission Information requested by the TTC. That, found Madam Justice Kiteley, made Gottardo's bid non-compliant.

Because Gottardo had also raised the defence of rescission, Madam Justice Kiteley decided to deal with it as an alternative. A sort of "what if".

> **Rescission**. The setting aside of a contract, as an equitable remedy. If the contract is so *rescinded*, it is treated as if it had never existed.

Justice Kiteley assumed that Contract A between the TTC and Gottardo was created at the moment Gottardo submitted its bid. At that point in time, Gottardo had made a mistake but did not know it. The error was not such as to be obvious to the TTC. Contract A was formed.

Before the TTC accepted the bid and therefore before Contract B was created, both the offeror (Gottardo) and offeree (TTC) came to know that the offeror had made a mistake.

Counsel for the TTC argued that it was not an appropriate case for the court to exercise its discretion to invoke the remedy of rescission for the following reasons:

**148** BIDDING AND TENDERING: WHAT IS THE LAW?

(a) Upholding the terms of Contract A would not produce a result that is unjust or unconscionable because:

(i) Gottardo is a sophisticated contracting company, with an office staff of approximately 10 people, and a full-time Estimator with 30 years of experience in preparing tenders, who, with the help of his staff, prepared the tender in this case;

(ii) Gottardo posted a bid bond issued by CGU;

(iii) the plaintiff and defendants all are commercial entities each of which contemplated the possibility of Gottardo being selected as the successful bidder, and then refusing to enter in Contract B, and allocated the risk associated with this eventuality through the bid bond;

(iv) Gottardo and its Estimator clearly understood that they would be bound by the terms of their bid as submitted, including the irrevocability of the price;

(v) Gottardo and its Estimator chose to develop a process whereby the Form of Tender was signed in blank in advance then filled in at the last minute following advice given by telephone. This system was adopted for the commercial advantage of Gottardo and Gottardo cannot now look to the courts, through the doctrine of equity, to make it whole from the commercial consequences of its decision;

(b) Contract A was formed at the moment of submission of the bid at which time the mistake was not known to either the contractor or the owner;

(c) There was no fraud, misrepresentation or other act on the part of the TTC which induced or contributed to the mistake on the part of Gottardo;

(d) The public bid process would become a sham if contractors were permitted to withdraw bids within the period during which the price was irrevocable.

Justice Kiteley, however, found that it would be unfair and unjust to enforce Contract A, and gave the following reasons:

(a) This case is not fundamentally about the integrity of the public bidding process. Rather, it is fundamentally about the interpretation of the Instructions to Tenderers the drafting of which was wholly in the control of the TTC.

(b) I have found that Gottardo made an honest and inadvertent mistake. It is not a situation, as suggested by Lee, of a contractor saying: oh guess what, I don't want this for whatever reason. Gottardo had a logical explanation.

(c) From December 20th when Gottardo's letter was received by the TTC, officials rejected his request out of hand. There is no evidence that any consideration was given by TTC that perhaps its own contractual language distinguished it from the legal position on which it relied. Instead, on the recommendation of its staff, TTC "snapped" at it;

(d) The evidence of Gottardo is consistent with what he said in his December 20th letter, namely that if the TTC took action against the bid bond, it would "cause severe and irreparable financial hardship". Mr. Gillott took the position that that evidence ought not to be accepted without corroboration. However, there was no cross-examination on that part of Gottardo's evidence. There is no reason not to accept it and I do;

(e) The "commercial advantage" argument appears to be based on the decision of the trial judge in Ron Engineering where he described the "brinkmanship with the sub-trades", a practice "obviously" for the benefit of the bidder The Supreme Court of Canada did not comment on that argument. I agree that the bidder reaps some advantage from the "brinkmanship with the sub-trades". But it is ultimately the owner who benefits from the highly competitive bidding which is generated in the typical public bid process, the rules for which are established by the owner. The fact that Gottardo may have achieved some "commercial advantage" over its competitors does not mean that it is disentitled to equitable relief;

(f) there was no fraud or misrepresentation or other act on the part of the TTC which induced or contributed to the mistake on the part of Gottardo. But as pointed out above, the threshold is not the conduct of the offeree. The threshold is the effect on the offeror.

In this alternative scenario, concluded Justice Kiteley, if Contract A was created, the TTC was not entitled to enforce it. Gottardo was entitled to rescission.

---

*It should come as no surprise that Madam Justice Kiteley's decision has been appealed. There is a lot of meat for the Ontario Court of Appeal to chew on.*

*The central finding of Her Honour was simple enough. Since it takes a compliant bid to create Contract A, none was created because Gottardo's bid was non-compliant. The TTC had created — perhaps inadvertently — a two-stage closing. That gave each bidder two opportunities to be compliant, or not. All bids had to comply with the Tender Documents and that included a timely and correct second submission — in the form of the Tender Submission Information. In baseball, one out of two is batting 500. But, in a two-stage bid submission, the result is a non-compliant bid.*

*Two-stage bid processes are fairly common. Given the mayhem at most bid closings, contractors prefer the extra time to identify the trades they wish to carry and to flesh out supplementary information such as unit prices. However, most owners who use a two-stage process do not open the envelope carrying the first stage (the bid price) until the envelope carrying the second has been received. Then, if the second envelope doesn't arrive, the first is returned unopened.*

*Her Honour's alternative holding that Gottardo would have been entitled to rescission of Contract A had it come into existence is troubling. Contract A was established in the first place to avoid and prevent just what Gottardo tried to do — withdraw its bid after it realized it had made a mistake.*

When Her Honour found that it would be "unfair and unjust" to enforce Contract A, she rejected the argument of the TTC that Gottardo had submitted its bid knowing that this situation could develop and accepting the risk if it did. This is a risk which bidding contractors have freely accepted since Ron Engineering closed the invisible mistake exit ramp some 23 years ago. It will be interesting to see what the Court of Appeal says about the application of rescission to the very situation that Contract A was invented to address.

## SUMMARY

The following points apply to mistaken bids:

- A bidder will be liable for damages for breach of Contract A if it withdraws its irrevocable bid, except where the bid has a serious (material) error "on the face of the bid".
- A latent price error by the low compliant bidder cannot be amended and accepted even if the amended bid remains the low compliant bid.
- Where bids are submitted in two stages, the assesment for compliance applies to both submission stages.
- The bidder submitting a non-compliant bid may rely on its own non-compliance to avoid Contract A.
- If a subcontractor who has been "carried" has made a mistake in its bid to the general contractor, the subcontractor, bound by Contract A, is in the same position with the general contractor as the contractor with the owner.

# Chapter 7

# NEGLIGENCE AND MISREPRESENTATION

*Ron Engineering sent the traditional principles of common law covering the basics of bidding and tendering spinning. However, if you are looking for compensation in a bidding dispute, there are other ways to "skin the cat" which have nothing to do with contract.*

*A breach of contract, as we have seen, makes the person who made the breach liable to the other contractual party for damages.*

*Another way a person can become liable for damages is through the commission of a tort — that is, a breach of a duty imposed by common law.*

*A person can be guilty of a breach of contract only if there is a contractual relationship with another party. The law of torts knows no such barriers: liability is for careless conduct in general, regardless of contract.*

*In the context of bidding, we are primarily concerned with damage or financial loss caused by negligence, which is a breach of a duty of care.*

*We started the first chapter of this book with a classic case illustrating the principles of contract law. A proper exposition of negligence in general, and negligent misrepresentation in particular, is impossible without at least the basic outline of another classic: the Hedley Byrne case, a decision of the House of Lords in England.*

## HEDLEY BYRNE & CO. LTD. v. HELLER & PARTNERS LTD.

House of Lords; June 1963

Until comparatively recently, there could be no liability in law for negligence unless it resulted in physical damage or injury.

The only person who could sue for negligence causing *only* economic (that is, financial) loss was a party in a contractual relationship — for example, a client could sue his or her architect if he or she suffered economic loss due to the architect's negligence, but the contractor on the job could not sue because the contractor had no contract with the architect. The *Hedley Byrne* case changed that and triggered a revolution even greater than that caused by the *Ron Engineering* decision.

Hedley Byrne & Co. Ltd., an advertising firm, requested a reference from Heller & Partners Ltd., a bank, regarding a client. Heller supplied a favourable reference without properly checking. Hedley Byrne relied on the reference and extended credit to the client, who promptly went into bankruptcy. Hedley Byrne lost a lot of money and sued Heller for negligently misrepresenting the client's financial situation.

The defence maintained that the reference was given to Hedley Byrne's bankers without a fee and without any knowledge of what it would be used for. Heller knew nothing of Hedley Byrne and never intended the information to be communicated to them. The information was only an expression of opinion and entailed no liability:

> When a person sets up as an investment adviser, that imposes on him a duty to the persons he advises. But no duty is raised simply by asking a person for advice. By asking a policeman the way one does not make him one's traffic adviser.

Heller argued further that if there were liability, it would be only to the bank that actually made the enquiry; it could not be taken over by third parties like Hedley Byrne. Heller owed them no duty. Furthermore, its loss was only economic, without any physical damage or injury.

Hedley Byrne, however, maintained that, contract or no contract, fee or no fee, there should be no distinction between a negligent act resulting in physical injury and a negligent word:

> It would be strange if a person who handled his pen so carelessly as to put out X's eye were liable to pay damages, but not if he handled it so carelessly in writing that X was financially ruined.

The judge at the trial held that the bankers were negligent but agreed with the defence that it owed no duty of care to Hedley Byrne. The Court of Appeal likewise held that there was no duty of care, and it was, therefore, unnecessary to consider whether the finding of negligence was correct.

In the House of Lords, Lord Morris of Borth-y-Gest summed up the decision of the House:

> It should now be regarded as settled that if someone having a special skill undertakes, with or without a contract or remuneration, to apply that skill for the assistance of another person who relies on such skill, a duty of care will arise.

The Heller bank knew that what it said would be passed on to some unnamed person — a customer of the bank — making the inquiry. In these circumstances, Heller owed a duty of care towards the unnamed person, whoever it was, and was liable for not exercising such care in giving the reference. The bank need not have answered the inquiry, but since it did, it was obliged to exercise care.

Unfortunately for Hedley Byrne, the bank had added a *disclaimer of liability* to the reference, saying, in effect: "We accept no responsibility for our statement." Thus, even though the bank was guilty of negligent misrepresentation, Hedley Byrne won no damages. In view of the disclaimer, it was not reasonable for Hedley Byrne to rely on the reference. If it chose to receive and act upon Heller's reference, it could not disregard the definite terms under which the reference was given. It could not accept a reply given with a disclaimer of liability and then reject the disclaimer.

In a more recent case, *Queen v. Cognos*, the Supreme Court of Canada agreed with the *Hedley Byrne* decision and laid down the following five requirements that a plaintiff must establish to collect damages for negligent misrepresentation. They are as follows:

- There must be a duty of care based on a "special relationship" of closeness or proximity between the person making the representation (the representor) and the person to whom the representation is made (the representee).
- The representation in question must be untrue, inaccurate or misleading — in other words, it must be a misrepresentation.
- The representor must have acted negligently in making the misrepresentation.
- The representee must have relied on the negligent misrepresentation, and his or her reliance must have been reasonable in the circumstances.
- The representee must have suffered damages as a result of his or her reliance on the misrepresentation.

The crucial lessons for the construction industry contained in *Hedley Byrne* and *Cognos* are neatly summed up by the authors of *The Canadian Law of Architecture and Engineering*, Beverley McLachlin (now Chief Justice of Canada) and Wilfred Wallace (formerly of the Court of Appeal of British Columbia):

> Architects and engineers frequently make statements and representations that certain materials are suitable for certain purposes; that their designs are adequate; that the structure to be built complies with local bylaws; and that subsoil conditions are of a certain character.
>
> Such statements cannot be made casually to persons who may rely upon them. Care must be taken to ensure they are accurate, and if doubt exists, a disclaimer of accuracy should be made.

*The following cases illustrate how the principles of liability for negligence in general, and negligent misrepresentation in particular, have been applied in bidding and tendering situations.*

## CARMAN CONSTRUCTION LTD. v. CANADIAN PACIFIC RAILWAY CO.

Supreme Court of Canada; September 1982

This case illustrates the two main tools available to a contractor that suffers losses due to incorrect information supplied by an employee of the owner, or in a similar situation: one is based on contract law, the other on the principles of tort.

On the other hand, the case also shows how an owner can successfully protect itself by using a carefully drafted disclaimer clause in the contract and bid documents.

On September 6, 1977, Carman Construction Ltd. received a bid package from the Canadian Pacific Railway Company (CPR) for rock excavation on a railway project in Ontario. The package contained a letter of invitation to bid, printed Instructions to Bidders, a blank proposal form and a specimen form of contract — but no information as to the quantity of rock to be removed.

The period of time available to Carman to prepare its bid was extremely short: the deadline was barely three days later, on the morning of September 9.

Carman visited the site and took some measurements. Its vice-president, Mr. Fielding, went to the CPR offices to discuss its bid with someone in the engineering department. Fielding could not remember later who this man was, but Fielding told him that, based on the information available, Carman was unable to submit a price. The man volunteered a figure of 7,000 to 7,500 cubic yards.

Carman prepared and submitted its bid on the assumption that 7,500 cubic yards of rock were to be removed. The contract was for an upset price of $109,260 for work and material. The price included a fee of approximately 20% for profit, which amounted to $18,200. CPR accepted Carman's bid, and the parties signed a formal contract.

Clause 3 of that contract had the contractor declare that he had examined the site conditions "and all other matters which can in any way affect the work" and that he did not rely on "any information given or statement made to him in relation to the work by the company".

Clause 5 had the contractor guarantee the upset price. Even if the cost of the work exceeded the upset price, that was all the contractor would be paid. If the work cost less, the contractor would be entitled only to the cost of the work, plus the fee.

Carman completed the work and received the contract price. It then surveyed the job and found that it had removed 11,043 cubic yards. Carman submitted a claim for $32,282 plus 20% for its "overrun" fee. CPR refused to pay, and Carman took it to court.

The trial judge dismissed the action. He felt that the ingredients were there to support a claim:

- in contract, for breach of a *collateral warranty*; and
- in tort, for *negligent misrepresentation*.

However, he concluded that the exemption clause (Clause 3) precluded Carman from winning. The Court of Appeal dismissed Carman's appeal. Finally, all nine judges of the Supreme Court of Canada took a look at the claim.

A collateral warranty is a contract running side by side with the principal agreement. The collateral contract, which is often verbal, has the effect to vary or add to the terms of the principal contract and, therefore, the law views it with suspicion.

Justice Martland, who wrote the decision, quoted Lord Moulton in the English case *Heilbut, Symons & Co. v. Buckleton*:

> Any laxity on these points would enable parties to escape from the full performance of the obligations of contracts unquestionably entered into by them and more especially would have the effect of lessening the authority of written contracts by verbal collateral agreements relating to the same subject matter.

For this reason, not only the terms of the collateral warranty but also its existence must be established, as in the case of any other contract, by proving that the parties, in fact, intended to enter into such a contract.

The court found no evidence that the statement made by the CPR employee regarding the expected quantities of rock was intended as a collateral warranty. The existence of Clause 3 destroyed any chance of such a warranty.

Fielding knew about Clause 3 before he visited the CPR office. He knew that if Carman's bid were accepted, the contract would contain this clause. Indeed, the clause was

> ... clearly meant for the sole purpose of ensuring that, if prospective bidders got any information on this subject from [CPR's] employees, they would be relying upon it at their own risk.

There was an additional ground for denying the existence of a collateral warranty. Such a warranty, if it existed, would contradict the express terms of the principal contract, namely Clause 3. A collateral agreement cannot be established if it is inconsistent with, or contradicts, the written agreement.

Carman's second line of attack followed the principles laid down in *Hedley Byrne & Co. Ltd. v. Heller & Partners Ltd.* Those principles can be summarized as follows.

A reasonable person, knowing that somebody else was putting trust in him or her or relying on his or her skill and judgment, has three possible courses of action:

- He or she can keep silent or decline to give the information or advice that the other person is seeking.
- He or she can give an answer with a clear qualification that he accepts no responsibility for it, or that it was given without the preparation or research that a careful answer would require.
- He or she can answer without any such qualification.

If he or she chooses to adopt the last course, he or she accepts a legal duty to exercise such care as the circumstances require in making the reply. This is called the "duty of care". If he or she fails to exercise that care and causes damage, he or she is liable for negligence.

The trial judge dealt with Carman's claim on the basis that the contractor had established negligent misrepresentation but that Clause 3 was a disclaimer clause which exempted CPR from liability.

Justice Martland preferred to regard that clause as a statement by CPR that it did not assume any duty of care. No claim in negligence can succeed if there is no duty of care.

This was not a case in which, after making the contractor a negligent misrepresentation in order to induce it to enter into a contract, the owner afterwards inserts an exemption clause into the contract in order to insulate itself against liability for the misrepresentation.

When Fielding asked a CPR employee for information, he knew that if he obtained an answer he would be relying on it at his own risk. Carman submitted its bid, knowing that it would have to assume the risk of relying on any information obtained from the owner's employees. The Supreme Court, therefore, dismissed Carman's claim based on negligent misrepresentation.

---

*The following case was described by the trial judge as "a problem with no solution". There certainly was a problem, but a reasonable approach by the owner could have resulted in a solution rather than a lawsuit.*

# MAWSON GAGE ASSOCIATES LTD. v. R.

Federal Court of Canada; October 1987

Mawson Gage Associates Ltd., an electrical contractor, submitted a bid to Cana Construction for the electrical work on a project undertaken by the Department of Public Works (DPW). Mawson Gage's bid was incorporated into Cana's bid to DPW, and the company was named as the electrical subcontractor.

Cana's bid was successful, and this was communicated to Mawson Gage. However, shortly after learning of its successful bid, Mawson Gage discovered that a section of the specifications relating to "Growth Chamber Controls" had been omitted from the set of bid documents which it had obtained from the owner. Mawson Gage immediately informed Cana, DPW and the project architect.

Almost a month later, Cana received official notification that it had been awarded the contract at the bid price. The award letter made no mention of the Mawson Gage problem.

Mawson Gage's officers attempted repeatedly to have their problem resolved, but with little success. They expressed their reluctance to sign a subcontract that included work for which they had been unable to quote a price. Cana's response was that it had to sign a contract with DPW at the bid price, or it would forfeit its bid bond; therefore, Cana expected the subcontractor to sign as well.

If that did not happen, Cana would award the electrical subcontract to the next lowest bidder and sue Mawson Gage for the difference in cost. Cana's position was that the problem was between Mawson Gage and DPW.

Having to choose between signing the subcontract or facing a lawsuit, Mawson Gage chose to sign.

DPW declined to accept responsibility for the missing specification. DPW's engineer wrote to Mawson Gage:

> Public Works Canada has entered into a contract with Cana Construction Co. Ltd. Upon entering into this contract, the general contractor has stated that the total work in the contract documents will be carried out for the tendered price. This of course includes the section of work in question. If there is any dispute by Mawson Gage Associates Ltd., I suggest that it be dealt with between the contracting parties, not with Public Works Canada.

The above suggestion is a form of dispute resolution known as "let me hold your coat". In the end, DPW could not avoid being a combatant.

DPW faced a dilemma in attempting to find a resolution for the subcontractor's problem when it was first discovered. Even if DPW had acknowledged its responsibility for the defects, the path from that admission to a proper solution, short of cancelling the whole bid process, would have been far from clear.

Special concessions for the subcontractor within the bidding process were contemplated, but it was difficult to envisage how they could be implemented without offending all the unsuccessful electrical subcontractors.

If the electrical subcontract had been changed, it would presumably have required a corresponding amendment to the general contract. That could hardly have been done without reconsidering the unsuccessful general contractors.

In the words of the judge at the subsequent trial, the discovery of the error in the plans created a problem that seemed to have no solution.

In these circumstances, Mawson Gage completed the entire electrical subcontract. It was not paid for the work on the growth chamber controls so it sued for negligent misrepresentation and breach of contract arising from a defective set of plans and specifications.

Liability for negligent misrepresentation will be found if the following criteria are found to exist (a somewhat modified restatement of the *Queen v. Cognos* principles):

- a duty of care;
- a negligent misrepresentation or the equivalent, as in this case, a negligent omission to convey necessary information;
- reliance on the misrepresentation or lack of information;
- reliance by the injured party that was reasonable under the circumstances; and
- loss resulting from the reliance.

Mawson Gage cited the decision in *Walter Cabott Construction Ltd. v. R.* In that case, Justice Mahoney said:

> I have no difficulty in finding that the relationship between the person who invites tenders on a building contract and those who accept that invitation is such a particular relationship as to impose a duty of care upon that person so as to render actionable an innocent but negligent misrepresentation in the information which he conveys to those whom he intends to act upon it.

The court found that DPW had a duty of care to the subcontractor; that DPW delivered to Mawson Gage deficient plans and specifications; and that Mawson Gage's erroneous bid resulted from relying on DPW's misrepresentation.

Finally, Mawson Gage had to prove that it had suffered a loss as a result of its reliance on the plans and specifications supplied by DPW. The subcontractor showed that it was forced to complete the subcontract, including work on which it had not bid, for the originally tendered price.

DPW contended that Mawson Gage was not forced to do this, but did so by choice, and, therefore, the loss resulted not from the misrepresentation but from the subcontractor's decision to go ahead and perform the work at a price it knew did not cover all the work to be done.

In answer, the subcontractor used *Ron Engineering* to defend itself. It argued that its bid was irrevocable if accepted within 30 days. If it had withdrawn its bid or refused to perform the contract, the general contractor would have sued it — and probably won. The court agreed that it was reasonable, if not absolutely necessary, for Mawson Gage to sign the subcontract and get on with the work.

Another aspect of the case influenced the court's decision. If the subcontractor were unsuccessful, DPW would have received the electrical work without paying for it. Furthermore, that unjust enrichment would have resulted from an error committed by DPW. Such a result would be unjust.

For all these reasons, the court found DPW liable for damages resulting from its negligent misrepresentation.

---

*In the course of his reasons, the trial judge called this case "a problem with no solution". In that statement, he seemed to be referring to the predicament created by the bid documents. However, there was a solution to the problem had the owner been sensitive to it. In reality, the owner had a bid from the contractor which carried Mawson Gage that was not responsive to the bid documents. When the truth came out, the bid was a mistake albeit an invisible one.*

*What the owner could have done (as the owner in Ron Engineering could have done) was let the contractor withdraw as it requested. Another alternative was to invoke the privilege clause, reject the bid and go to the second bidder. That approach was not in fashion then, and it does not seem to be now. If the owner had rejected the bid, it probably would have spent less (of our) money in the long haul.*

## ST. LAWRENCE CEMENT INC. v. FARRY GRADING & EXCAVATING LIMITED

Supreme Court of Ontario; February 1989

St. Lawrence Cement Inc., operating as Dufferin Construction Company, was in the process of preparing its bid for the construction of a highway when it received a telephone quotation from Farry Grading & Excavating Limited, a subcontractor experienced in the business of excavation and earth removal.

Dufferin carried Farry's unit prices in its bid documents and was awarded the contract for the project. On the same day, Dufferin notified Farry by telephone that it wanted to proceed immediately with the subcontract. Farry, however, appeared unwilling to take action.

During subsequent meetings, Farry attempted to negotiate an increase in its unit prices because it was unable to make use of the original dump site that did not require any dumping fees. Alternative locations would mean substantial fees. The problem was not resolved, and, finally, Farry informed the contractor that it would not proceed with the subcontract.

Dufferin awarded the subcontract to another firm at increased cost and took Farry to court to recover damages resulting from (a) its alleged breach of bidding contract; and (b) negligent misrepresentation in quoting the unit prices.

Could Dufferin maintain concurrent actions at the same time: one for breach of contract and the other for the *tort of negligence*?

> **Tort**. A wrongful act or omission other than breach of contract (such as negligence, trespass or nuisance) for which damages can be obtained in a civil court by the person wronged.

Following the 1986 decision of the Supreme Court of Canada in *Central Trust Company v. Rafuse* this became possible. Dufferin was free to pursue both options and, in the end, choose which of the two provided the most advantageous result.

Was there a contract between Dufferin and Farry? Was it reasonable for Dufferin to rely on the telephone quote from Farry? The court found that, considering the nature and strategy of the bidding process, it was reasonable for a contractor to rely on last-minute telephone quotes. Accordingly, there was a bidding contract (Contract A) between the two parties; it was breached by Farry, and this resulted in damages to Dufferin.

The court found that the amount of damages for breach of contract was the increase in the price Dufferin had to pay the new subcontractor as the result of Farry's default — a total of $143,484.

The court then turned its attention to Dufferin's claim based on the tort of negligent misrepresentation. The essential elements of this tort were set out in the 1964 decision of the House of Lords in *Hedley Byrne v. Heller and Partners*. To succeed Dufferin had to prove:

- that it reasonably relied on the advice of Farry;
- that Farry had a particular skill or knowledge; and
- that Dufferin's reliance on Farry's advice caused Dufferin to suffer damage.

Dufferin was a general contractor for road works. Farry, a haulage subcontractor, quoted certain prices to the contractor shortly before Dufferin submitted its bid on the main contract. The court found that this was a communication made in a business context by a party with special knowledge (Farry) to another party (Dufferin) who, by the very nature of the bidding process (known to Farry), would reasonably rely on that communication as a contractual commitment.

Thus the elements of the tort were present. Farry's defence that Dufferin ought to have known that Farry's quote was unrealistically low and, therefore, not to be taken seriously was dismissed by the court because Farry had special knowledge and expertise in the field of haulage and was offering its services on that basis. Dufferin was not obliged to verify bids because part of what it was contracting for was an accurate prediction of costs by the specialist subcontractor.

The purpose of damages in tort is to put the plaintiff in the same position it was in before the harm happened. This meant compensating Dufferin for its out-of-pocket loss, which was the difference between the amount tendered by Farry and the amount paid by Dufferin to the new subcontractor.

This happened to be the same as the measure of damages for the breach of contract, namely $143,484 and that was the amount Dufferin was awarded. Of course, Dufferin could not recover twice: it would only receive damages in *either* its contract *or* its tort action — in this case, it did not matter which, since the amount was the same either way.

---

*The following decision contains a very detailed discussion of what constitutes negligent misrepresentation, in all its various manifestations, and a number of other legal issues. At approximately 300 pages, the decision is almost a textbook of construction law courtesy of the learned trial judge who wrote it and, of course, the litigants who jointly financed much of the associated cost.*

## OPRON CONSTRUCTION CO. LTD. v. R.

Court of Queen's Bench of Alberta; March 1994

Between 1981 and 1985, Opron Construction Co. Ltd. worked for the Alberta Department of the Environment on a dam project. It encountered several soil condition problems which resulted in a claim against the department for breach of contract and, concurrently, for deceit and negligent misrepresentation. Justice Feehan found that for various sections of the project:

- the department possessed a great deal of relevant information which it did not disclose to the bidders;

- some information was deceitfully concealed from the bidders; and
- some soil conditions were negligently misrepresented.

He also found the department in breach of its contract with Opron.

Opron suffered substantial losses. However, a great many of them were Opron's fault. Justice Feehan found that Opron's bidding estimates and its execution of the work were done in a negligent manner, therefore, Opron was "the author of a substantial portion of its losses".

The trial lasted 14 months and resulted in a decision that is well over 300 pages long. For obvious reasons, only the gist of the legal analysis decision will be reported here, not the complicated details of the project.

The bid package contained, in legal terms, *representations* made by the owner to the bidders. These representations may be:

- made expressly, in writing or orally; or
- implied either from an express statement or from acts or conduct.

An *implied* misrepresentation is just as actionable as an express one. Express representations of fact can be found in plans and drawings. For example, in *Cardinal*, a Bell installation was misrepresented on a plan, and the contractor suffered damage. The engineer and the owner were found liable for negligent misrepresentation.

A statement as to the contents of a document is also a representation. In 1881, in *Arkwright v. Newbold*, Lord James said:

> Supposing you state a thing partially, you may make a false statement as much as if you misstated it altogether. Every word may be true, but if you leave out something which qualifies it, you may make a false statement.

An estimate can also constitute a representation of fact. In *K.R.M. Construction Ltd. v. B.C. Railway*, the owner's estimate of approximate quantities was extremely inaccurate.

The trial judge accepted expert evidence that the words "estimate" and "approximate" in the construction industry were generally understood to indicate a maximum variation of 20%. The B.C. Court of Appeal upheld this conclusion:

> [The trial judge] concluded that the way in which the invitation to tender was presented was intended to convey to prospective bidders that sufficient and proper engineering procedures had been followed in arriving at the estimated quantities....
>
> He held that in setting forth this information in the invitation to tender [the owner] was representing that it had adequate information available and had used such information to arrive at the approximate quantities contained in the schedules.

Thus an inaccurate estimate may be seen by the court as a misrepresentation of fact.

Liability for misrepresentation can also spring from silence in certain circumstances, or from a failure to disclose certain facts. In *Leeson v. Darlow* the court noted:

> Active concealment of a fact is equivalent to a positive statement that the fact does not exist... for example, to cover over the defects of an article sold with intent that they shall not be discovered by the buyer has the same effect in law as a statement in words that those defects do not exist.

The Supreme Court of Canada in *Banque de Montréal v. Hydro-Québec* treated the duty to inform in a contractual context as an obligation arising from the primary obligation of good faith between the contracting parties. Justice Gonthier said:

> ... the obligation to inform and the duty not to give false information may be seen as two sides of the same coin.

While that decision was based on the Civil Code of Québec, the same idea is also found in the common law: once a party has embarked on imparting information to the other, he or she must ensure that the information is not only accurate, but also complete. Justice Gonthier added:

> The owner's obligation to inform increases with its expertise relative to the contractor's, particularly when it provides information to the contractor which falls within its field of expertise, and that information is incorrect.

The owner cannot evade liability by asserting that it was merely repeating the representation provided by its consultants. Those preparing bid packages generally possess much greater knowledge about the circumstances of the project than the bidders do. The bidders also may lack the opportunity or time to acquire necessary information — as the Supreme Court of Canada found in *Edgeworth Construction*. This differential in knowledge imposes a duty of care on those preparing the bid documents to ensure that they are accurate.

In order to be entitled to damages, the bidder must prove that in submitting its bid and entering the contract, it relied on the representations in the bid documents. Obviously, if the bidder did know that the representations were incorrect, it cannot assert that it had been misled by them.

The contractor is not required to investigate whether the representations are true or not, even if informative sources are available. In 1880, in the case *Redgrave v. Hurd*, the court said:

> If a man is induced to enter into a contract by a false representation, it is not a sufficient answer to him to say, "If you had used due diligence you would have found out that the statement was untrue. You had the means afforded you of discovering its falsity, and did not choose to avail yourself of them."

The common law does not place on the contractor the burden to investigate the truth of the representations in the bid package independently, as the contractor is, in principle, entitled to rely on them.

Opron submitted that the following two terms should be implied in its contract with the department:

- There were no facts within the department's knowledge which had not been disclosed to the bidders.
- The information provided by the department in the bid documents had been furnished in good faith, in the honest and reasonable belief that it was complete and accurate.

Terms can be implied in a contract only if they are not in conflict with any express terms or disclaimer clauses. For example, in *Catre Industries Ltd. v. R.*, the contractor tried to imply in its contract a term obliging the owner "to provide soils information of reasonable accuracy".

But Catre's contract contained two clauses that were in conflict with the proposed implied term. These clauses stated:

> No responsibility will be assumed by the [owner] for the correctness or completeness of the data shown....
>
> The bidder is required to investigate and satisfy himself of everything and of every condition affecting the works....

This was enough to destroy Catre's implied term.

With respect to soil conditions, Opron's contract stated, among other disclaimers, that the accuracy of this information was not guaranteed. However, the implied clauses proposed by Opron were different from Catre's: Opron did not rely on a guarantee of the accuracy of the soils information, as Catre did, but merely assumed that the owner did not possess any information contradicting the soils information in the bid package. There was no disclaimer clause in the contract addressing that question.

There is a general legal principle that if there is no clause forbidding reliance, a bidder is entitled and expected to rely on the bid documents as conveying the best information the owner or its engineer can give. An implied contractual term that the owner does not possess any information contradicting the bid documents "lies within the boundaries of this principle", decided Justice Feehan.

This, in itself, was not enough for the court to admit the proposed implied terms. Opron also had to show that the implied terms were legally necessary. Opron argued that to permit the owner to disclose information which it knows is contradicted by other information would be to subvert the contract itself. It was, therefore, necessary to imply the terms it proposed.

Justice Feehan agreed. He noted that the implied terms did not place on the owner a general legal obligation to provide prospective bidders site information. Rather, the terms required an owner who has already supplied bidders some information to disclose *all* relevant information in its possession. The judge, therefore, accepted the terms submitted by Opron as implied in its contract.

In the last few years, the Supreme Court of Canada has taken steps toward a judicial recognition of a doctrine of good faith in contractual relations. In *Lac Minerals Ltd. v. International Corona Resources Ltd.*, Justice La Forest said:

> The institution of bargaining in good faith is one that is worthy of legal protection in those circumstances where that protection accords with the expectations of the parties.

Justice Feehan applied this to the Opron case:

> I conclude that in the circumstances of this case there is a covenant implied by law that the parties will deal fairly and in good faith with one another in the exchange of information.
>
> It is reasonable, where the owner or its agents impart critical information in the tender documents which form part of the contract, that there is an implied covenant that such information has been furnished in good faith, in the honest and reasonable belief that it is complete and accurate, with all material information provided, in the sense that there is no inconsistent information within the owner's knowledge bearing upon the tender or the performance of the contract.

The department had an obligation of good faith and fair dealing to Opron. Therefore, it should have disclosed to Opron that it possessed material soils information which was inconsistent with, or which contradicted, the information contained in the bid documents. The department breached this obligation.

Did the department act deceitfully in withholding some of the information? In order to sustain an action in deceit or fraud, said Justice Feehan, it is not enough to establish a misrepresentation: "neither bungling, ineptitude nor gross negligence establishes fraud."

The classic case dealing with this issue happened in 1889: in *Derry v. Peek*, the House of Lords held that the test of deceit is whether the person making the representation honestly believes his or her statement to be true.

A statement made recklessly is an instance of lack of belief, "for one who makes a statement under such circumstances can have no real belief in the truth of what he states".

Similarly, if a person making a false statement shuts his or her eyes to the facts, or purposely abstains from inquiring into them, this is a sign that honest belief is absent. He or she need not actually know that the statement is false: it is enough that he or she suspects it, or that he or she neglects to inquire whether it is accurate.

It is also immaterial, with some exceptions, why the false statement was made. The finding of deceit does not depend on an *intention* to deceive. The key question is whether the person making the statement intends that the person to whom he or she made the statement should *rely* on it.

Justice Feehan was satisfied that the misrepresentations and non-disclosures made by the department with respect to certain areas of the project were fraudulent and deceitful:

- The department actually knew of the misrepresentations and non-disclosures, or, at least, did not honestly believe that its representations were true.
- The department intended that Opron should rely on the contract documents as being all of the documentation necessary to enable Opron to prepare a proper bid.

The judge found that the department misrepresented the conditions in some areas of the project, but the misrepresentations were not made deceitfully. Were they made negligently?

The special relationship of proximity or "neighbourhood" existing between an owner and bidder gives rise to a duty of care. This legal duty is to exercise reasonable care when preparing and presenting bid documents, and when disclosing information regarding the site conditions, the nature of the work required and so on.

It is not enough that the owner believes that what is contained in the bid documents is accurate. It must not be negligent in forming and expressing that belief, or in deciding what information to withhold, or in inquiring as to the existence and relevance of other information affecting the work. In determining the standard of care to which the owner must adhere, Justice Feehan listed the following general principles:

- When preparing bid documents, the average bidder must be kept in mind, not one that is unusually cautious, conservative or has special knowledge or experience.
- When preparing a bid, a bidder is entitled and expected to rely on the bid documents as conveying the best information the owner can give.
- Where an owner has special knowledge gained through years of investigation and then invites bids within a relatively short bid period so that the bidder must rely on the bid documents and has little chance to verify the information independently, the owner must disclose all facts which may reasonably have a bearing on the bids.

- If the owner has not verified information in the bid documents, it has a legal duty to inform bidders in clear terms that they should not rely on this information and should be allowed adequate time for independent investigation.
- If there is no clearly worded clause forbidding reliance, and the bid period is short, the inference is strong that the bidders will place greater reliance on the bid documents.
- If, after the bid documents are prepared and delivered, the owner learns that they are inaccurate, it owes a legal duty to the bidders to inform them of the true facts.

The court found that the department had not met the standard of a reasonable owner issuing an invitation for bids, and that it was, therefore, negligent.

Having found that the department had been negligent and in breach of its contract with Opron, the court had to check if the department's liability was limited or negated by the express terms of the contract in its various disclaimer clauses. As a matter of general principle:

- any clause purporting to limit or exclude liability must be expressed clearly and will be limited in its effect to the *narrow meaning* of the words employed;
- such a clause must clearly cover the *exact circumstances* which have arisen;
- if there is any ambiguity, the rule of *contra proferentem* will be applied, that is, the clause will be interpreted *against the party which drafted it*; and
- no limitation clause can excuse fraud.

The Instruction to Tenderers prepared by the department stated:

> The tenderer is required to investigate and satisfy himself of everything and of every condition affecting the works to be performed and the labour and material to be provided....

The law, said Justice Feehan, holds that the bidder will have discharged the obligation to "investigate and satisfy himself" if it accepts the basic information in the bid documents, unless the contract:

- specifies other sources of information; or
- provides that in making the investigations the bidder must not rely on the material supplied by the owner or its consultants.

He quoted, as particularly instructive in this regard, the 1972 decision of the Australian High Court in *Morrison-Knudsen v. Australia* and the 1914 decision of the Supreme Court of the United States in *Hollerbach v. United States*. In the latter, the court said:

> We think it would be going quite too far to interpret the general language of the other paragraphs [of the contract] as requiring independent investigation of facts which the specifications furnished by the government as a basis of the contract left in no doubt....
>
> In its positive assertion of the nature of this much of the work [the government] made a representation upon which the [bidders] had a right to rely without an investigation to prove its falsity.

*Hollerbach* was adopted in *Cardinal Construction* to support the conclusion that the contractor in that case was entitled to rely on negligent misrepresentations in the bid documents, even though the documents contained a clause instructing the contractor to investigate for itself.

Evidence was clear that in Opron's case, a visual site inspection could not have revealed the true subsurface conditions, nor were the bidders allowed enough time to conduct their own intensive soil investigations. The court held that Opron had discharged the obligation placed on it by the "investigation clause" by reviewing and accepting the basic information contained in the bid documents and by making a site visit.

Section 1.13 of the contract disclaimed responsibility for the accuracy of any information in the bid documents, using the following words:

> The accuracy of this information is not guaranteed and in no circumstances whatever shall the Engineer be held responsible or liable to the Contractor for any consequences resulting from the strata, groundwater and/or conditions as actually encountered being different from those implied in this information.

The court interpreted this disclaimer as meaning merely that the engineer is not the insurer or guarantor of the accuracy of the information.

The clause did not negate the representation that the engineer had prepared the technical studies in accordance with sufficient and proper engineering standards and procedures, or that the owner had disclosed in the bid package all relevant geotechnical information in its possession.

In any case, noted Justice Feehan, section 1.13 purported to shield the engineer from liability, not its employer, the department. He concluded that the disclaimer clauses in Opron's contract did not limit or exclude the department's liability.

## BG CHECO INTERNATIONAL LIMITED v. B.C. HYDRO AND POWER AUTHORITY

Supreme Court of Canada; January 1993

In 1982, B.C. Hydro called for bids for the construction of two transmission lines in British Columbia. Before submitting its bid, BG Checo International Limited sent a representative to inspect the area by helicopter. He noted that the clearing of the right-of-way was still going on and assumed that it would be completed. The bid documents provided that the clearing would be done by B.C. Hydro and would not be the responsibility of the contractor. In February 1983, BG Checo was awarded the contract. However, no further clearing ever took place. The contractor found logs and debris strewn all over the site and, as a consequence, suffered delay and additional costs.

BG Checo sued B.C. Hydro for negligent misrepresentation and breach of contract. During the trial, documents produced by B.C. Hydro showed that it had been aware all along that there was a problem with the clearing and had known the impact this would have on the contractor. As a result, BG Checo amended its statement of claim to include *fraudulent misrepresentation*.

B.C. Hydro's project manager explained that "one of the principles of specifications is not to belabour specific things for fear of being accused of leaving something else unknown to the contractor because you didn't highlight it". He insisted that it was unnecessary to include any words of warning in the contract because the logs on the right-of-way would be obvious to a bidder viewing the site.

The trial judge, however, saw this as a form of "tender by ambush" because the representation made with regard to the condition of the right-of-way was false, and that B.C. Hydro made it knowing that it was false.

B.C. Hydro further pointed to a clause in the contract which provided: "it shall be the tenderer's responsibility to inform himself of all aspects of the work", while another clause required that "the contractor shall inspect and examine the site and shall satisfy himself as to the nature of the site ..." and so on.

B.C. Hydro could not rely on such clauses, replied the judge, to shift the risk to the contractor when B.C. Hydro intended that the contractor should act on the false representation.

B.C. Hydro then pleaded that the contractor did not show reasonable care in its investigation of the site. Even if true, countered the court, this would not change B.C. Hydro's liability to the contractor for fraud. The contractor was not bound to mitigate its loss, as it was not until after the project was completed that it first learned of the fraud perpetrated by B.C. Hydro.

The question of a defence for fraud based on a lack of reasonable care was clearly disposed of in the case of *United Services Funds v. Richardson Greenshields of Canada Limited,* where the court said:

> Once the plaintiff knows of the fraud he must mitigate his loss but until he knows of it, in my view, no issue of reasonable care or anything resembling it arises at law. And, in my opinion, a good thing, too. There may be greater dangers to civilized society than endemic dishonesty. But I can think of nothing which will contribute to dishonesty more than a rule of law which requires us all to be on perpetual guard against rogues lest we be faced with a defence of 'Ha, ha, your own fault I fooled you'. Such a defence should not be countenanced from a rogue.

The court was satisfied that BG Checo would not have submitted its bid on the basis it did if B.C. Hydro had disclosed the true facts relating to the condition of the right-of-way. Therefore, B.C. Hydro was bound to compensate the contractor for the damages of approximately $2.6 million suffered as a result of being fraudulently induced to enter into this contract, plus costs and interest. B.C. Hydro appealed.

The B.C. Court of Appeal found no evidence of an intention to deceive and, therefore, rejected fraud, but decided that B.C. Hydro was guilty of negligent misrepresentation which induced BG Checo to enter into the contract. It awarded BG Checo $1.1 million for the misrepresentation and sent the question of breach of contract, and damages for the breach, back to trial. The parties appealed to the Supreme Court of Canada.

The Supreme Court had to reconcile the inconsistencies of the contract. On the one hand, there was the specific provision obliging B.C. Hydro to clear the right-of-way. On the other, there were the general provisions placing the responsibility to check everything about the site on BG Checo before bidding. The contract also said that if any errors or ambiguities appeared in the bid documents, it was BG Checo's responsibility to have those clarified before submitting its bid.

The cardinal rule the courts follow when interpreting a contract is that the various parts of the contract must be interpreted in the context of the intentions of the parties as they appear from the contract taken as a whole.

When there is an inconsistency, the courts attempt to find an interpretation which can give meaning to each of the conflicting terms. If there appears to be a conflict between a general term of contract and a specific term, the courts may infer that the parties had intended the scope of the general term to be such as to exclude the specific term, or, on the contrary, that the specific term should override the general.

The court decided that the contract specifically required B.C. Hydro to clear the site, and that duty was not negated by the more general clauses relating to errors and misunderstandings in bidding, site conditions and contingencies.

The court had to decide next whether the fact that BG Checo had a contract with B.C. Hydro prevented the contractor from concurrently suing B.C. Hydro in tort for negligent misrepresentation. In general, where a party who suffers damage appears to have a cause of action in contract and in tort, it has the right to sue in either or both unless limited by the parties themselves, through their contractual arrangements.

The contract between BG Checo and B.C. Hydro did not delete B.C. Hydro's common law duty not to negligently misrepresent that it would have the site cleared by others. Therefore, concluded the court, BG Checo was permitted to sue in tort for negligent misrepresentation.

The measure of damages for breach of contract is different from that of the tort of negligent misrepresentation.

*Contract*: The plaintiff is to be put in the position it would have been in had the contract been performed as agreed.
*Tort*: The plaintiff is to be put in the position it would have been in had the misrepresentation not been made.

In situations of concurrent liability in tort and contract, however, it would seem anomalous to award a different level of damages for what is essentially the same wrong on the sole basis of the form of action chosen.

The factor that can make a difference is the element of *bargain*. A contract may represent a good or a bad bargain for either party. Damages in contract normally take this into account and include compensation for the loss of a good bargain. Tort damages do not.

This is why it was important to decide what BG Checo would have done had it known the true facts regarding the site. If the court concluded that, but for the misrepresentation, the parties *would not* have entered into a contract, the court would have to deny "loss of bargain" damages.

If the parties *would* have entered into a contract anyway, the contractor should be awarded damages for loss of a good bargain, even if unrelated to misrepresentation, and the amount of damages in tort and in contract should be very similar. Of course, under no circumstances would the contractor be able to get compensation for losses caused by its own poor performance or by market forces that are a normal part of business transactions.

The trial judge had found that there would have been no contract. The Court of Appeal, however, decided that BG Checo would have entered into a contract with B.C. Hydro — but at a higher price. It decided that the increase would have equalled the cost of the extra work made necessary by the improperly cleared site, plus a 15% margin for overhead and profit.

The Supreme Court agreed that BG Checo would have entered into the contract in any case. Evidence at trial showed that BG Checo was keen to break into the B.C. market. Therefore, BG Checo was entitled to be compensated for all reasonably foreseeable losses caused by B.C. Hydro's negligence. It was foreseeable that BG Checo would lose money while clearing the site. It was, therefore, entitled to the cost of such extra work plus overhead and profit.

But to compensate only for the direct costs of clearing is to suggest that the only tort was the failure to clear, said the judges. The real fault was that B.C. Hydro *misrepresented* the situation. That misrepresentation might have caused other losses. For example, it might be that in having to devote its resources to that extra work, BG Checo failed to meet its original schedule and incurred acceleration costs.

The Supreme Court decided that the matter should be referred back to the trial judge to determine whether there were such indirect losses and whether they were the foreseeable result of the misrepresentation.

The court agreed that all BG Checo's claims for breach of contract should also be sent to the trial judge. On the claim for breach of contract BG Checo would have to be reimbursed for all expenses incurred as a result of the breach, whether expected or unexpected.

The only exception would be expenses that may have been so unexpected that they were too remote to be compensable for breach of contract. This restriction corresponded, to all practical purposes, to the test of reasonable foreseeability applicable in tort. Thus, the damages in contract would include not only the costs flowing directly from the improperly cleared site but also indirect costs such as acceleration costs due to delays.

---

*The road to the resolution of the dispute between BG Checo and B.C. Hydro was costly and long; it took almost ten years between the award of the contract and the final decision. Worst of all, the Supreme Court of Canada decided that a major part of the dispute would have to go back to the trial judge. The parties thus ended up at square one.*

*As a consequence, the owner made a business decision: It took a hard look at its construction contract and made significant revisions to some of the harsh clauses which used to place the burden of unexpected site conditions squarely on the shoulders of the contractor.*

*The two following cases seemed to indicate that the law of bidding and tendering might veer off into a new direction — **negligence** — leaving the intrica-*

cies of Contract A/Contract B behind. As the Ontario judge said in the first of the two cases (emphasis added):

> I want to emphasize my view that to apply the approach of [**negligence**] to all disputes which arise out of relationships created by the tendering or bid depository process can only bring clarity to the **somewhat befuddled** [Contract A/Contract B] approach with which I have struggled over these last six months.

His decision was quickly followed by a British Columbia judge, who was even more creative. The B.C. judge abandoned the "befuddled" Contract A/Contract B approach to find liability but, nevertheless, used a "unilateral contract" of the Ron Engineering type to assess damages.

Both cases then went to their respective Courts of Appeal ...

## TWIN CITY MECHANICAL v. BRADSIL (1967) LIMITED

Court of Appeal for Ontario; November 1998

The decision of Justice Dandie in *Twin City Mechanical v. Bradsil* created quite a stir when it was released in December 1996.

The facts were as follows. In 1990, the Province of Ontario called for tenders for the construction of a laboratory building. The bidding was to be done through the local bid depository. Twin City, which submitted a bid of $9,540,000, was the low bidder for the mechanical portion of the job. The prime contract was awarded to Bradsil (1967) Limited, which submitted the lowest bid at $23,322,000.

Bradsil knew in advance that Twin City was a non-union company and, therefore, could not do the subcontract work. It used this knowledge to shop the mechanical work after it had already won the prime contract and eventually awarded the mechanical work to another bidder for $340,000 less than Twin City's bid.

Justice Dandie found as a fact that the apparent attempts of the general contractor to negotiate a construction contract with Twin City were not carried out in good faith.

He imposed on the Province, and any other owner calling for bids through the bid depository system, a duty of care to protect a subtrade from the improper conduct of the general contractor. He further imposed on a public body or institution a higher duty of care to assure that everyone who participated in the process was treated fairly and in accordance with the "good faith" principle.

The trial judge also accepted the opinion of an expert, which had been called by Twin City, that the owner ought to have declared Bradsil's bid "informal", since it was conditional on the implied condition that the subcontractor enter into a union agreement. The judge decided that the Province, as the tender-calling authority, had a duty to adjudicate the labour compatibility issue between its general contractor and subcontractor.

The Province appealed. The Court of Appeal did not find it necessary to decide whether the Province owed subcontractors such as Twin City a duty of care to ensure the integrity of the bid depository process. Assuming that such a duty did indeed exist, the court decided that in taking the steps that it did, the Province had complied with this duty.

Contrary to the view expressed by the trial judge, the Province was not required to investigate or adjudicate the complex legal and factual issues arising from the dispute between Bradsil and Twin City, nor was it obliged to mediate the dispute.

The Province required Bradsil to work out an agreement with Twin City. If Bradsil could not do so, it was allowed to replace Twin City with one of the subcontractors that had bid on the project at a price not less than that submitted by Twin City. This, decided the Court of Appeal, was a reasonable response designed to preserve the integrity of the bid depository process.

The trial judge had also held that the Province was negligent because it had failed to ensure that Bradsil complied with the condition that the new subcontract price not be less than Twin City's. The Court of Appeal disagreed on that point too.

It would have been preferable for the Province to check up on Bradsil, but the Province was not obliged to do so. It was entitled to rely on Bradsil's unqualified assurance that the new price would not be less than Twin City's. To the extent that Bradsil breached this condition, the Province may have been justified under the contract in reducing the amount it was obliged to pay Bradsil. This, however, was a matter between the Province and Bradsil — it had nothing to do with Twin City.

Finally, the court reviewed the trial judge's finding that Bradsil's initial bid was informal and should, therefore, have been disqualified.

On its face, the bid was not informal. Bradsil's ploy was only brought to light after the closing date for bids had passed. To say that the Province should, at that point, have disqualified the tender as informal would have required the Province to become embroiled in the legal and factual dispute between Bradsil and Twin City. However, the Province was not required to do so.

The Court of Appeal decided that there was no basis on which the Province could be found liable to Twin City; therefore, it allowed the Province's appeal. Twin City's claim was dismissed, and the Province was awarded its costs, both at trial and on appeal.

CHAPTER 7: NEGLIGENCE AND MISREPRESENTATION  **175**

# KEN TOBY LTD. v. B.C. BUILDINGS CORPORATION

Court of Appeal for British Columbia; April 1999

Like the Ontario decision in *Twin City,* the following British Columbia case took relations of the various parties in the bidding process outside the scheme set out in *Ron Engineering* and introduced the tort of *negligence* into the mix.

Three months after the Court of Appeal for Ontario reversed *Twin City,* the B.C. Court of Appeal was asked to review the trial decision in *Ken Toby*.

The facts of *Ken Toby* were somewhat unique. In 1995, B.C. Buildings Corporation (BCBC) called for bids for renovations to the Royal British Columbia Museum in Victoria. One of the requirements was that subtrade bids had to be submitted through the bid depository of the B.C. Construction Association.

Ken Toby Ltd. submitted bids for the Unit Masonry and Granite and Marble sections of the work. Its bids to four general contractors carried a price of $497,378 for both sections separately, and the same price for both sections combined. Toby submitted a higher price ($506,666) to a fifth general contractor, Wigmar Construction Ltd., but again the amount was the same for both sections, separately and combined.

Obviously, Toby did not wish to undertake the comparatively small Unit Masonry work unless it was also awarded Granite and Marble. Toby's distaste for this work alone was shared by other bidders; they simply did not bid on that section.

In the end, Toby was the only subcontractor to submit a bid for Unit Masonry. One other bid was received for Granite and Marble, at a price of $272,665.

Paragraph III(4)(c)(iii) of the bid depository Rules of Procedure stated that

> ... unless otherwise stipulated in the specifications, General Contractors receiving a single Bid Depository bid must use that bid.

The bid depository closed on March 21, 1995. About ten days after the closing, a general contractor called BCBC's Director of Project Management and informed him (incorrectly) that the bid (or bids) for Unit Masonry were over $200,000. BCBC had estimated the cost of that section to be no more than $15,000.

In response, the director instructed the project architect to issue an addendum (Addendum 14) substituting a cash allowance for Unit Masonry. However, this clashed with Rule III(1)(B) of the bid depository which stipulated:

> To avoid occasions where Addenda are published at too late a date for Bidders to be notified through the medium of plan bulletins and other industry publications, the Tender Calling Authorities should not publish Addenda less than three (3) working days prior to closing in the case of trades covered by the Bid Depository.

General bids closed on March 24. BCBC accepted the lowest bid of Wigmar Construction Ltd., which later named Marchesi Marblecraft Ltd. as the subtrade for Granite and Marble.

Toby sued BCBC, claiming that if Addendum 14 had not been issued, its bid for Unit Masonry would have been used by Wigmar, since it was the only bid for that section.

Justice Burnyeat, the trial judge, was in large measure influenced by the judgment of the Ontario Court in *Twin City*. He decided that:

- when Toby submitted its bid to the bid depository, a contract was created between BCBC and Toby (Contract A);
- when both parties agreed to be bound by the bid depository rules, BCBC accepted the duty to act in good faith to any subcontractor who submitted bids to general contractors through the bid depository;
- when BCBC issued Addendum 14, it was in breach of its *contract* with Toby; and
- when BCBC failed to follow the rules, it was in breach of its duty of care to deal with Toby in good faith, and, thus, liable for the tort of *negligence*.

The judge awarded Toby damages of $129,491, which represented Toby's economic loss resulting from the breach of contract and the breach of the duty to bargain in good faith.

The B.C. Court of Appeal had to decide two questions:

- What contractual relationship, if any, existed between BCBC and Toby?
- What duty of care, if any, did BCBC owe Toby ?

Justice Goldie, writing for a unanimous court, rejected the idea that any legally enforceable relationship between BCBC and Toby ever arose.

The trial judge had identified the contractual relationship between BCBC and Toby as a unilateral contract of the Contract A type.

The first difficulty Justice Goldie had with the trial judge's finding of a unilateral contract was that BCBC's invitation to bid — its offer, according to *Ron Engineering* — was not made to subcontractors such as Toby. It was an offer addressed to the general contractors.

It was literally impossible for a subcontractor to respond to BCBC's invitation. A subcontractor could not comply with the fundamental condition that the bid must include a price for the entire project; this was Justice Goldie's second difficulty with the trial decision. A basic condition of a unilateral contract is that the performance of the contract must meet the precise conditions of the offer. Obviously, if somebody offers a reward for his or her lost dog — a typical unilateral contract — you cannot claim the reward if you bring a cat. Hence, there is an offer but no acceptance.

The judge added:

> It may be argued these are technical objections. That may be, but I know of no policy reason supporting a contractual relationship in the circumstances I have sketched. Freedom of contract plays a major role in the commercial world. Where parties bargain to define rights and obligations in a chosen relationship, the courts ought not to be astute in stretching the concept of the unilateral contract to cover the absence of offer and acceptance.

The court concluded that no contractual relationship between Toby and BCBC was formed when Toby deposited its bids before the closing of the bid depository.

The second issue considered by the court was negligence and the duty of care. The logic of the trial judge's finding of liability for negligence is apparent from the following quotations:

> The step taken by [BCBC] in issuing Appendix 14 was in breach of Rule III(1)(B)...
>
> In acting in contravention of the Rules of Procedure and in issuing Addendum 14 without making proper enquiry as to whether the single bid on Section 04210 (Unit Masonry) was, in fact, in excess of $200,000.00, [BCBC] breached duties owed to [Toby].
>
> [BCBC] insisted upon the use of the Bid Depository for the use of General Contractors and Subcontractors but they then specifically targeted the sole Bid of [Toby] when they issued Addendum 14....
>
> [This] ... deprived [Toby] of its rightful expectation arising out of the Rules that it had to be named by all General Contractors as the Subcontractor on Section 04210 (Unit Masonry) or would likely be named as the Subcontractor on the combined Sections 04210 (Unit Masonry) and 04424 (Granite and Marble).
>
> [BCBC] owed all users of the Bids Depository System including the plaintiff a duty of care to ensure that everyone who participated in the process which it had initiated and which it had insisted upon was treated fairly in accordance with the good faith principle and in accordance with the Rules of the Bids Depository System. ...

Justice Goldie, however, reasoned differently: The owner's primary duty to all subcontractors is to take reasonable steps to ensure the integrity of the bid depository process. In the judge's view, Addendum 14 was consistent with this duty.

The trial judge took the wording of Rule III(1)(B) as mandatory. He relied partly on the evidence of expert witnesses. But such witnesses had no business interpreting the bid depository rules in light of industry custom.

Where there is no bid depository system, as was the case in *Chinook Aggregates*, the court may accept expert evidence regarding industry custom and usage. If such custom and usage is certain and well known and universally accepted within the industry, the court may treat it as an implied term of a written contract.

But expert evidence was not required in the *Ken Toby* case, since the bid depository rules already reflected construction industry custom and usage. Rather, the question was the meaning of Rule III(1)(B) in its context, and this was for the court to decide.

Justice Goldie concluded that the word "should" in the rule was not mandatory. Its purpose was simply to facilitate the smooth working of the system.

Addendum 14 was in response to information received by BCBC after the closing date for subtrade bids but before the deadline for the general contractors. The information indicated that the lowest bid for Unit Masonry was over $200,000 for work that was worth no more than $15,000. Thus, there was a need to prevent a possible misuse of the bid depository system.

Toby's bid was not genuine, in the sense that it was not intended to be the lowest bid. It was intended to prevent acceptance of a bid for Unit Masonry alone.

Other bidders that did not wish to be awarded that section alone did not provide a bid price. Their abstention was consistent with the bid depository system, but Toby's bid was not, said the judge. Therefore, in his view, BCBC's response was appropriate.

This alone would dispose of the finding that Addendum 14 was issued in bad faith. However, the matter did not end there.

During the trial, it came to light that on March 22, unknown to BCBC, one of the general contractor bidders obtained a quotation from Toby for Unit Masonry alone. The quotation appeared in a fax, which stated in part:

Our Price: $12,350.00 + G.S.T.

Toby withdrew this quote the next day, after it found out it was the only bidder for that section. The trial judge took a lenient view of this episode and concluded that the withdrawal of the quote left the bids filed in the bid depository intact. The Court of Appeal considered Toby's action a breach of bid depository rules.

There was one further ground for denying Toby's claim. In the official bid form used by Toby, the following sentence appeared immediately above the signature block:

> We have read the Rules of the Bid Depository and agree to be bound by all of those Rules, including the Exclusion of Liability clause contained in them, which we understand precludes any claim by us relating in any way to the operation of the Bid Depository.

The fact that Toby had signed this release represented the final blow to its claim. The Court of Appeal unanimously allowed BCBC's appeal and dismissed Toby's action with costs.

Ken Toby (and Twin City — see the previous case above) applied for leave to appeal to the Supreme Court of Canada. Both leave applications were dismissed on February 17, 2000.

---

*The most likely situation where an owner finds itself liable in negligence to a contractor or subcontractor is where there is a significant error or omission in the bid documents. Whether that shortcoming is incorrect or incomplete information does not matter. All parties expect that the bid documents will be relied upon to found the bids that are made by suppliers, subtrades and general contractors. This creates the "special relationship" which gives rise to an obligation that the issuer of bid documents is not to put these parties in harm's way. As long as the error or omission is not obvious, parties who rely on such bid documents will obtain relief.*

*In both Twin City and Ken Toby, the appellate courts refused to use tort theory to award damages on whatever relationship/duty of care may exist between an owner and a subtrade. But, both appellate courts examined the behaviour of the owners within an envelope that recognized that there is some relationship and therefore some duty. Because it turned out that neither owner had acted outside this duty, there was no liability. It would be unwise for owners to assume an air of total indifference toward trades by contractors or by their own actions.*

*No question, concurrent liability in contract and tort has a natural and appropriate place in the field of bidding.*

## SUMMARY

The following conclusions can be drawn from the cases in this section:

- An owner may be liable to a subtrade in negligence where the owner's error lies in the very division of work in which that trade is involved and where the trade is damaged in the process.

- Remedies in contract and negligence may be pursued concurrently in a bidding situation.
- A party that reveals only part of the story can be liable in negligence in the same way that the party can be negligent if it provides an incorrect story.
- Where bid documents require a bidder to perform due diligence, the extent of the investigation required depends on the amount of time the bidder has to make the enquiry and depends on the cost or difficulty of the enquiry process.
- Disclaimer clauses in bid documents, as in all other contract documents, will be strictly construed against the party that attempts to rely on them.
- A duty of care owed by an owner to a subtrade can arise when there is no contractual relationship between the two, but that duty is unlikely to oblige the owner to take extraordinary measures to protect the subtrade from a contractor that fails to comply with the bid documents.
- A carefully drafted disclaimer clause may enable the owner to shift some risk to the contractor, but the economic sense of such an action is questionable.

# Chapter 8

# DUTIES OF CONSULTANTS

*The duties of consultants are, of course, legion. Books continue to be written on the subject. Here we will take only a brief look at the duties of consultants with regard to the process of bidding and tendering.*

*The first four cases reviewed in this chapter reflect disputes that occurred after Contract B was entered. They are instructive because all of the cases reached back to the bid documents to determine whether the consultant had given the bidders the full picture.*

## CARDINAL CONSTRUCTION LTD. v. BROCKVILLE (CITY)
Supreme Court of Ontario; February 1984

In 1978 Cardinal Construction Limited entered into an agreement with the City of Brockville to reconstruct sewers and water mains under a street. The bid documents on which Cardinal had successfully bid for the work were prepared by Kostuch Engineering Limited.

The documents showed a Bell installation under the street and described it as underground cable. Cardinal based its bid on this information. When the work started, the cable turned out to be a concrete-encased duct running the length of the street. The result was a substantial increase in the cost to the contractor. When negotiations failed to settle the matter, Cardinal took the City and the engineer to court.

The contract between Cardinal and the City imposed two obligations on the parties if the character of the work changed from that on which the contractor had bid, either because of a written order of the engineer or because of incorrect written information supplied by the City. The obligations are as follows:

- The contractor must proceed with the work.
- The city, if requested, is obliged to negotiate compensation for the resulting increase in cost to the contractor.

Was the contract provision applicable? As decided by Justice Henry:

The character of the work is to be determined from the contract documents; anything that deviates from the information supplied and that affects the contractor's production or cost in a significant way is a change in the character of the work.

Since Cardinal relied on the information in the bid documents and bid on the assumption that the Bell installation was a flexible cable only to find a rigid concrete structure, this changed the character of the work as a result of incorrect information supplied by the City. In particular, the rescheduling of the work, made necessary by the presence of the concrete duct, transformed the work from a quick autumn job into one requiring a winter shut-down with its attendant costs, to be resumed and completed in the spring.

The City took the position that it was unnecessary to identify the installation as a concrete duct. The contractor could have found that out from other information supplied in the drawings, as some bidders, reading between the lines, in fact did. The court could not accept this. The following statement of Justice Henry has often been quoted (emphasis added):

> The documents must be prepared for tender having in mind the *average bidder*, not the bidder who has special knowledge or experience. Among bidders there will be some who are unusually cautious and conservative ...; others who have specialized knowledge of Bell installations ...; others who have general knowledge and experience ...; and others who may be bidding on this class of project with little or no experience.
>
> Given the general principle that a bidder is entitled and expected to rely on the tender documents as conveying the best information the engineer can give, it is not good enough in my opinion to provide information that is misleading, incomplete or inaccurate with the intention that the more experienced or knowledgeable bidder will ferret out the problems from 'clues'. The information should be clear and intelligible to all bidders and in this case it was not.

It is interesting to note in passing that the cautious experienced bidder that made the enquiries and knew about the concrete duct did not get the contract — its bid was too high.

Counsel for the City argued that the character of the work never changed: all the factors affecting the project, such as subsurface conditions and the nature and location of utilities, were present before bidding. The contractor's investigation simply did not go far enough, it failed to appreciate the true character of the work as it actually was.

General conditions of the contract required the contractor to declare that it had investigated the character of the work to be done for itself or, alternatively, if the contractor had not investigated, that it would assume the risk of unforeseen expense such as might be caused by changed soil conditions.

The contractor also had to declare that in tendering, it would not rely on information furnished by the City about, among other things,

... the general or local conditions and all other matters which could in any way affect the performance of the work under the contract other than the information furnished in writing for or in connection with the tender or the contract by the engineer.

This (in the words of the judge) "convoluted and confusing provision" was interpreted by the court, as the principles of law demand, as *contra proferentem* — against the party who drafted it. It was taken to mean that while Cardinal may not have fully investigated the character of the work, it still had the right to rely on the written information furnished by the engineer in the bid documents. Thus the disclaimer clause did not let Brockville off the hook.

Counsel for the City claimed that the custom or practice of the trade required that Cardinal should further investigate the Bell "cable" as an implied term of the contract. The court disagreed:

I am unable to find that there was any practice or understanding in the trade that a contractor bidding would investigate information supplied by the engineer to verify its accuracy. He would be expected to do so, however, if the nature of the utility was not specified, or if the documents raised some question in his mind as to the correct interpretation of the information necessary to his bid.

As an alternative to its claim for additional compensation, Cardinal claimed damages against the City by reason of *negligent misrepresentation* regarding the nature of the Bell installation, which induced the contractor to enter into the contract. As another alternative, Cardinal tried breach of *collateral warranty*.

In circulating the bid documents, the City intended to induce bidders to bid and to commit themselves to perform the work for a price, while relying on the information contained in the documents. In these circumstances, the law imposed a duty on the City to take care that the information was accurate, or to warn bidders that they should verify it.

The court found that the City had breached that duty of care: the representation as to the nature of the Bell installation was negligently made by the City through its agent, Kostuch Engineering Ltd. Moreover, Kostuch and the City became aware of the error before bids were submitted. They had a duty to convey this information to bidders, but failed to do so.

Finally, Cardinal made a further alternative claim for damages against the City for breach of an implied term of the contract: the owner must give the contractor uninterrupted and exclusive possession of the site. The court concluded that, indeed, such a term was implied in the contract, and that it was breached by the City when its engineer, following intervention by Bell, suspended the work for the winter.

Against the possibility that it might not recover its claim from the City in full, Cardinal started an independent action against the engineer Kostuch. Since Cardinal did not have a contract with Kostuch, the action was framed in tort of negligence and based on the principles laid down in the decision of the House of Lords in *Hedley Byrne & Co. v. Heller & Partners*. The principles were restated by Justice Henry and applied to the *Cardinal* case as follows:

- The engineer that prepares bid documents owes a duty of care to bidders that are known to rely on the information contained in the documents.
- The engineer's duty is to exercise reasonable care that the information presented reflects with reasonable accuracy the nature of the work.
- If the engineer has not verified specific information, the engineer has a duty to inform bidders in clear terms that the engineer does not vouch for its accuracy and that each bidder must investigate for itself.
- If the engineer learns during the bid period that some of the information supplied is incorrect, the engineer has a duty to inform bidders of the true facts.

The court found that Kostuch had breached its duty of care to Cardinal as a bidder. Cardinal was, therefore, entitled to damages from Kostuch. Of course, the damages from the engineer would be reduced by what Cardinal was paid by the City. The court was not asked to assess the amount of the claim. This was, by consent, left to a referee.

## BROWN & HUSTON LIMITED v. YORK (BOROUGH)

Ontario Court of Appeal; October 1985

Brown & Huston Limited was the successful bidder on a contract for the construction of an underground pumping station for the Borough of York, Ontario.

The bid documents had been prepared by York's consulting engineer, Fenco Consultants Ltd. Although the documents reproduced portions of a soils report prepared by Geocon, a soils consultant, they did not refer to the report itself, and they omitted important information concerning ground water levels.

The general contract signed by Brown & Huston obliged the contractor to satisfy itself as to all local *surface* conditions. The contractor chose a method of construction based on the information contained in the bid documents. This method proved to be entirely inappropriate in light of the high water table level encountered during construction. Construction, therefore, had to be completed using more expensive procedures.

Brown & Huston sued both York and Fenco for the additional costs. The trial judge found that the engineer owed the contractor a duty of care, as Fenco must have known that bidders would rely on the bid documents, particularly when the documents did not require contractors to satisfy themselves about *subsurface* conditions. The engineer was negligent in not providing the bidders necessary information, such as the soils report.

The judge determined that Fenco was 75% liable for the missing information, but Brown & Huston was also liable, to the extent of 25% of its damages, for failing to make inquiries concerning the report. Brown & Huston's engineer recognized that certain symbols on one of the bid documents indicated groundwater. Furthermore, the project was close to the Humber River, and the presence or absence of water was of vital importance to contractors.

The action against York was dismissed. The trial judge did not feel that the Borough was responsible for Fenco's negligence.

Both Fenco and Brown & Huston appealed to the Ontario Court of Appeal which held, in part, as follows:

- Fenco was negligent for failing to make the soils report available to bidders.
- As between Fenco and Brown & Huston, the allocation of 75% to 25% liability ought not to be disturbed.
- Under the construction contract, the legal obligation to supply the soils report for the information of bidders rested on York.

The court stated: "Clearly [York] had appointed Fenco its agent for this limited purpose and must be held liable for Fenco's negligence in failing to perform this function at least."

Accordingly, the Borough was held *jointly and severally liable* with Fenco for 75% of Brown & Huston's damages.

> **Joint and Several**. A legal expression meaning "together and separately". If two or more persons enter into an obligation that is joint and several, their liability for its breach can be enforced against them all or against any of them separately.

*In the next case, the Supreme Court of Canada finally confronted and, more or less, resolved a difficult question which, until now, had attracted different answers in different parts of the country: was the design professional liable to the contractor for errors and omissions in the bid package when there was no contractual link between them.*

## EDGEWORTH CONSTRUCTION LTD. v. N.D. LEA & ASSOCIATES LTD.

Supreme Court of Canada; September 1993

In 1977, the B.C. Ministry of Highways hired the engineering firm N.D. Lea & Associates to design approximately 20 km. of highway relocation. Edgeworth Construction Ltd. was awarded the construction contract. N.D. Lea was not named in the contract, and it had no further involvement in the project. The contract price was approximately $7 million, but through extra work orders, unit price changes and so on, the actual amount paid to Edgeworth by the ministry was just short of $20 million, yet the contractor claimed a further $22 million.

Edgeworth claimed its loss was caused by errors in the engineering design. It sued N.D. Lea and the individual engineers who worked on the project. The engineering firm and the employees applied for summary judgments to have both actions dismissed. The key argument they put forward was that in the circumstances, they owed no duty to the contractor.

The trial judge agreed. He held that Edgeworth's only recourse under its contract was against the ministry and dismissed the contractor's action. The contractor appealed.

"This is a claim for negligent misrepresentation or misstatement", said Justice Lambert, speaking for the Court of Appeal. He then discussed negligence, which, in the legal sense, is a breach of a duty of care. That duty arises:

- if it is foreseeable that the carelessness of the defendant will cause loss to a person who is put at risk by this lack of care;
- if the person who suffers loss justifiably and reasonably relies on the defendant's misrepresentation, and if the person whose negligent misrepresentation creates the risk can foresee that reliance; or
- if there is a relationship of *proximity* between the parties.

It does not matter whether those three concepts are treated as aspects of each other, or merely as being interrelated.

The most elusive of the three concepts is proximity. How should the court determine whether there is a relationship of sufficient proximity? The court decided that the web of contractual relationships in a construction project forms a crucial element, and that had to be considered in answering the question.

Edgeworth had a construction contract with the Ministry of Highways. When the ministry adopted N.D. Lea's plans and specifications in the bid package, those documents became the bidders' representations of the ministry. Under the contract, Edgeworth could properly recover its losses from the ministry.

By contrast, N.D. Lea had no direct relationship with Edgeworth — only with the ministry. The engineering firm had no opportunity to define the risks that it was prepared to assume, if any, in relation to Edgeworth. It was not compensated on the basis that it would be assuming such risks, nor could it influence the contract between the ministry and the contractor.

The contractor, on the other hand, had an opportunity to protect itself in the contract with the ministry and define the risks it was prepared to bear. It could reflect those risks in the bid price.

This meant, decided the court, that there was no relationship of proximity between N.D. Lea and Edgeworth which would impose on the engineer a duty of care to take all reasonable steps necessary to avoid causing the successful bidder economic loss. The court unanimously dismissed the contractor's appeal.

The matter did not end there. Late in 1993, 16 years after N.D. Lea first started work on the project, the Supreme Court of Canada rejected the reasoning of the Court of Appeal and allowed the contractor to proceed with its lawsuit against the engineering firm.

Madam Justice McLachlin wrote the decision for a unanimous court. Liability for negligent misrepresentation, said the judge, arises when a person makes a representation (such as a statement, a design or a specification), knowing that another may rely on it, and that other person, in fact, relies on the representation and suffers loss or damage. This was decided in 1964 by the House of Lords in the case *Hedley Byrne v. Heller & Partners* and never seriously challenged.

The facts alleged by Edgeworth met the *Hedley Byrne* test. N.D. Lea undertook to provide information (the bid package) for the bidders. The engineering firm knew that the bidders would use the information it supplied to prepare their bids. Edgeworth was one of the bidders. It reasonably relied on the bid package, and it alleged that it suffered loss as a consequence. These facts, decided the court, established a *prima facie* case against N.D. Lea.

> **Prima Facie.** Latin expression meaning "on the face of it". A prima facie case is one that has been supported by sufficient evidence for it to be taken as proved if the case is not rebutted on evidence acceptable to a court.

The courts below had taken the position that the ministry assumed all risks previously held by the engineers when it adopted the bid package and distributed it as its own. Not so, said Justice McLachlin. Indeed, the design representations, once made part of the bid package, became the representations of the ministry. But this did not mean that they ceased to be the representations put forward by

N.D. Lea. The contractor was relying on the accuracy of the engineer's design just as much after it entered into the contract with the ministry as before.

Finally, the engineering firm submitted that the court could deny on grounds of *policy* a duty of care which might otherwise arise. To allow the claim in tort against N.D. Lea to proceed would, as a general consequence, inhibit engineers and other design professionals from accepting limited fees for limited services (such as not providing site services).

This may be so, agreed the Supreme Court, but many professionals — accountants, for example — assume duties toward persons other than those with whom they have contracted, and they are held liable in tort if they perform such duties negligently. Typically, the additional risks are reflected in the fees that are charged in the contract. Alternatively, the professionals defend themselves by means of *disclaimers of liability*.

There was also an important policy consideration against the engineering firm. If it were to win its arguments, contractors would be obliged to do their own engineering. In the typically short period allowed for the filing of bids, the contractor would be obliged to do the work that took engineers months or even years. This would be an almost impossible task. Moreover, each bidder would be obliged to repeat a process already undertaken by the owner's consultant.

From an economic point of view, decided the court, it makes more sense for a consultant to do the engineering work and for the contractors to rely on that work — if there is no disclaimer or limitation of liability.

The risk of liability to third parties for design errors will be reflected in the cost of the engineer's services to the owner. But that is much better than requiring the owner to pay indirectly for the additional engineering which all bidders may be forced to do.

The position of the individual employees of N.D. Lea sued by Edgeworth was different, ruled Justice McLachlin. The only basis on which they were sued was the fact that each of them had affixed his or her seal to the design documents. This was insufficient in establishing a duty of care between the employees and the contractor. The seal simply attested that a qualified engineer had prepared the drawings — it did not guarantee their accuracy, nor was it sufficient to found liability for negligent misrepresentation.

The court, therefore, concluded that Edgeworth could proceed with its action against N.D. Lea but that the action against the individual engineers should be dismissed. So, some 16 years after the events that gave rise to the lawsuit, the contractor and the engineer were back where they started, assembling the documents and the witnesses required to establish and argue their respective cases before a trial judge.

CHAPTER 8: DUTIES OF CONSULTAN

*Design professionals did not need to be told by a court that there is a broad family of "persons" who may choose to rely on an engineer's work product. The Supreme Court of Canada extended that duty in time and to "persons" that the engineer had not expressly considered.*

*Disclaimer clauses are the means at hand for consulting engineers to avoid or at least limit their exposure. The most effective disclaimer is carried in the contract and binds the consultant's client. The challenge becomes extending protection afforded by privity of contract to those "persons" with whom the engineer has no contract. The difficulty lies in crafting a clause which recognizes the consultant's duty to its client (and may limit that in the bargain) but warns all other "persons" that the risk of reliance on the consultant's work product lies with them and not the consultant. Or, the consultant can attempt to get an indemnity from its client to cover claims like Edgeworth's. Will the indemnity be available? Good luck!*

*In the next case, the Supreme Court of Canada held that there were some limits to the duty of care the consultant owed to the contractor.*

## AUTO CONCRETE CURB LTD. v. SOUTH NATION RIVER CONSERVATION AUTHORITY

Supreme Court of Canada; September 1993

Auto Concrete Curb Ltd. submitted the lowest bid for the sixth stage of a contract for dredging the South Nation River, based on bid documents prepared by Kostuch Engineering Limited.

Contractors on the first five stages of the project used conventional land-based earth moving equipment — draglines and backhoes — for the removal of earth above and below the water line. Auto Concrete's bid was based on a less expensive method that employed a suction dredge and a settling lagoon.

The engineer advised the contractor that it had no objection in principle to the use of suction dredging "provided the necessary provincial environmental approvals are obtained".

Auto Concrete was unable to obtain the permits for doing the work as planned and was obliged to use the conventional method. In doing so, it lost considerable sums of money.

The contractor sued Kostuch, alleging that the engineer should have told it about the need to obtain permits and should have warned it that this might be difficult. The failure to do so, argued the contractor, constituted negligent misrepresentation on which the contractor relied and, as a consequence, it lost money.

The trial judge decided that the engineer should have foreseen the possibility of a contractor using suction dredging. The engineer should have made inquiries of the authorities and warned the bidders of the possible environmental restrictions. It was not reasonable, said the judge, to expect the contractor to make permit inquiries within two weeks after the award of the contract when the engineer had some four months during the preparation period of the bid documents.

The engineer should have known that the bidders would rely on the information in the bid documents, so its failure to include all relevant information constituted negligent misrepresentation. The judge found the River Authority vicariously liable for the negligence of the engineer because, in law, it was responsible for the actions of its consultants.

The Court of Appeal found no reason to interfere with the conclusions of the trial judge. The dispute was appealed to the Supreme Court of Canada.

Madam Justice McLachlin, speaking for the Supreme Court, held that the courts below had been in error. The standard of care imposed on an engineer preparing bid documents does not require it to advise bidders of the need to obtain permits to do the work where the method of executing that work is determined by the bidder itself.

It has long been established, she said, that the method of construction is entirely the contractor's responsibility. Neither the owner nor the design professionals employed by the owner have a duty to advise the contractor what method to choose or how to proceed once the owner makes its choice.

Applying this rule to Auto Concrete's claim, Justice McLachlin decided that Kostuch was not under a duty of care to advise the contractor about problems it might encounter in obtaining permits to do the work by the suction method as it had planned to do. Justices La Forest, Sopinka, Gonthier, Cory, Iacobucci and Major agreed. The Supreme Court of Canada allowed the engineer's appeal.

---

*The next case is remarkable because the unsuccessful lowest bidder took the unusual step of suing not only the owner but also the owner's consultant. The claim against the owner was for breach of the bidding contract (Contract A); the claim against the consultant was for the tort of **inducing** a breach of contract.*

# ACL HOLDINGS LTD. v. ST. JOSEPH'S HOSPITAL OF ESTEVAN

Saskatchewan Court of Queen's Bench; April 1996

ACL Holdings Ltd. was the lowest bidder for the construction of a hospital in Estevan, Saskatchewan. The owner, on the advice of its architect, Arnott Kelley O'Connor & Associates Ltd., awarded the contract to the second lowest bidder.

ACL then sued the owner for breach of Contract A. It claimed that it was an implied term of the bidding contract, based on industry custom, to award the contract to the lowest bidder. Furthermore, ACL submitted that the architect was also liable for damages because it induced the owner to breach the contract.

At the examinations for discovery, ACL admitted:

- that it is the architect's function to assess for the owner the qualifications of the bidders;
- that the architect, in its assessment of ACL, considered factors which it was entitled to consider; and
- that there was no indication of bad faith in the actions of either the owner or the architect.

Following these admissions, the architect applied for *summary judgment* to have ACL's claim against it dismissed.

---

**Summary Judgment.** Decision without a full and formal trial. A party to a civil action may obtain summary judgment when it can demonstrate that on the facts alleged by the claimant, there is no genuine issue for trial and it is almost certain that the claim would fail if a trial was held. The motion may be directed toward all or part of the claim or defence. In this way, a party, when faced with allegations delivered by an opponent who has no genuine claim or defence, can obtain a speedy judgment without going to trial. The application for summary judgment is heard by a judge *"in chambers"* rather than in the courtroom.

---

Legal authorities clearly state that summary proceedings must be used with great care. A party to a dispute should not be shut out from trial unless it is very clear that its action has no hope of success. Summary judgments will, therefore, not be granted when there is any serious conflict as to facts or points of law.

## Bidding and Tendering: What is the Law?

The tort of intentionally inducing a breach of contract was first recognized by an English court in 1853, in *Lumley v. Gye*. In order to succeed in such an action, a plaintiff must establish:

- that a valid and enforceable contract was in existence;
- that the defendant was aware of the contract's existence;
- that the defendant induced the breach of the contract;
- that there was wrongful interference; and
- that the plaintiff suffered damage as a result.

ACL's admission that it was the architect's job to assess bidders and that it based its assessment on the proper factors did not help the architect: they indicated that the architect did "interfere" with the bid contract.

The question whether the interference was wrongful is an independent element of the tort, said Justice Barclay. The court had to consider whether the architect's contractual and advisory relationship with the owner was a possible defence.

This question was considered in Alberta by Master Quinn, in the case *Spectra Architectural Group Ltd. v. Eldred Sollows Consulting Ltd.* Counsel in that case could refer the Master to only one Canadian precedent, so the Master based his decision mostly on American jurisprudence. In the United States, no tort of inducing breach of contract arises where the advisor has a duty, recognized by law, to give advice to a contracting party, and the contracting party has the right in law to terminate the contract based on the advice.

The only requirements for the existence of this privilege shielding the advisor are that his or her advice must be requested, must be given within the scope of the request, and must be honest. There is a fourth requirement where the advisor is under a duty to a third person to exercise reasonable care. If so, the advisor must observe that duty of care.

Clearly, said Justice Barclay, the tort of inducing breach of contract is not settled in Canada, but the law here recognizes the tort of *negligently* inducing breach of contract. The pleadings and admissions before him did not preclude a successful claim that the architect negligently induced St. Joseph's to breach the implied terms of the bid contract. The judge added:

> Furthermore, the matter under consideration in this application is whether [the architect] has shown, on clear and unequivocal admissions by ACL, that it is *impossible* for ACL's claim to succeed, not whether it is possible that [the architect's] defence has merit.

ACL's third admission that there was no bad faith did not assist the architect. ACL only claimed that the architect induced the owner to breach the bid contract, not that the architect did so in bad faith.

Since Arnott Kelley had not shown that it was impossible for ACL to succeed in its claim, the court denied the architect's application for an order dismissing ACL's claim.

---

*One must always be careful not to read too much or too little into a reported decision. In ACL Holdings, ACL failed. The court was not impressed with ACL's claims against the architect. That does not mean that an architect is free of any duty to a contractor upon whose bid it must report. Consultants should expect that their conduct in bid review will be held up against an objective standard of care established for such professional activities.*

*The next case deals with an all too common problem: unexpected soil conditions. If the contractor fails to get the compensation it believes is its due, a claim for negligent misrepresentation is bound to come.*

## WIGMAR CONSTRUCTION (B.C.) LTD. v. DEFENCE CONSTRUCTION (1951) LIMITED

Supreme Court of British Columbia; November 1997

In the spring of 1992, Defence Construction (1951) Limited called for bids for construction work at CFB Esquimalt, B.C. The following clause appeared in the bid documents:

Section 3.10 — Soils Information

1. Copies of soil information are available for inspection at the offices of Defence Construction (1951) Ltd., 300 Gorge Road West, Victoria, B.C. The soil report was prepared to design the project. Matters effecting construction may not have been addressed by the soils engineer. The contractor is fully responsible for matters affecting construction.
2. This soil data is for reference and information only and is not to be construed in any way as giving directions relative to the contract or as implying any guarantee as to working circumstances, soil performance, or subsurface condition.
3. The contractor remains fully responsible for assessment of soil and site conditions and shall take any additional steps necessary to assure himself of any such circumstances.

Wigmar Construction (B.C.) Ltd. was the successful bidder. It carried Den-Mar Equipment Rentals as subcontractor for excavation and rock removal.

DenMar encountered difficulties in the blasting, excavation and site preparation due to unexpected rock and soil conditions. Wigmar sued to recover the additional costs, claiming that Defence and its soils consultant negligently misrepresented the soil conditions, knowing that bidders, including Wigmar, would rely on the soil reports.

Wigmar pointed out that Defence approved 287 change orders, 10 of which were approved by the soils consultant. Wigmar took this as clear proof that the owner intended Wigmar to rely on the bid documents.

Wigmar also put the wording of the disclaimer clause under a microscope. It is not possible to separate design from construction, argued the contractor, so the soil report must have been prepared for construction even though the disclaimer said it was for design purposes only.

The disclaimer made the contractor fully responsible for "matters affecting construction". This means access to the site, deliveries, weather and the like, argued Wigmar, not fault lines in the rock. Furthermore, Wigmar's claim was for rock and not for soil, and, therefore, the words of the disclaimer which referred to "soil data" did not apply.

But Wigmar's principal point was that Defence and its soils consultant fully knew about the problems Wigmar encountered on the job before the contract went to bid. They knew or ought to have known of the fault lines and should have forewarned bidders. Instead, they withdrew the geotechnical reports so that they did not form part of the bid documents.

Defence countered that Wigmar did not rely on the representations in the geotechnical reports but rather on its subcontractor, DenMar. Wigmar expected DenMar to take the excavation to the appropriate elevations and was assured by the sub that this could be done for the subcontract price. Accordingly, Defence asked the court to dismiss the claim for negligent representations.

Justice Vickers relied on the decision of the Supreme Court of Canada in *Queen v. Cognos Inc.* There, the court listed the elements of negligent misrepresentation, already reproduced in the *Arrow Construction Products Ltd. v. Nova Scotia (Attorney General)* case and again in *Hedley Byrne & Co. Ltd. v. Heller & Partners Ltd.*.

Thus, central to the *Wigmar* case was proof of reliance on the geotechnical reports. For that, said the judge, Wigmar must show the following:

- Defence should have foreseen that Wigmar would rely on its representation.
- Wigmar's reliance was, in the circumstances, reasonable.

When is reliance reasonable? Again, the judge had a list of five general criteria handy:

- The representor has a direct or indirect financial interest in the transaction in respect of which the representation was made.
- The representor is a professional, or someone who possessed special skill, judgment or knowledge.
- The representation was provided in the course of the representor's business.
- The representation was given deliberately.
- The representation was made in response to a specific enquiry or request.

Armed with these lists, the judge gave Wigmar's claim short shrift.

Given the specific provisions of Article 3.10, it was not reasonable to expect that any bidder, and in particular Wigmar, would rely on the geotechnical reports. It could not be inferred from the approval of change orders that either Defence or its consultant intended Wigmar to rely on the reports.

On a plain reading of the contract documents, Article 3.10 was intended to apply to excavation work. It could not be restricted to the narrow limits of access or weather and the like.

Furthermore, it was evidently intended to apply to both rock and soil. The word "soil" included "rock", and that is how Wigmar understood it. Wigmar did not rely on the geotechnical reports in the preparation of its bid. It relied on DenMar to perform its contract.

"Whether the defendants knew or ought to have known the reports were inaccurate is not relevant to the issues in this case where there was no actual reliance on the reports by Wigmar", concluded Justice Vickers and dismissed Wigmar's claim.

---

*To paraphrase the policy arguments of the Supreme Court of Canada in Edgeworth: Should contractors be obliged to do their own soil tests in the typically short period allowed for the filing of bids? Does it make economic sense for each bidder to repeat tests already undertaken by the owner's consultant?*

*In Edgeworth, the court concluded that it is more reasonable for a consultant or owner to do the engineering and soil testing work, and for the contractors to rely on that work — if there is no disclaimer or limitation of liability.*

*In Wigmar, the disclaimer clause certainly did its job, but shifting the risk to the contractor in this way still makes no economic sense:*

- *Some sensible contractors, faced with the disclaimer clause that Wigmar found in the bid documents, probably decided not to bid, therefore, the owner had to choose from a smaller pool.*
- *Any sensible contractor that elected to bid most probably put in its bid a hefty contingency sum for unforeseen soil problems.*

- If no soil problems appear, the owner will have overpaid the contractor; if such problems do appear, the owner will have paid most of the costs by way of bid contingency or maybe even more than the contractor's costs.
- When there are soil problems, very few contractors will accept the extra costs without a major fight, contingency or no contingency; thus, the relations on the job tend to be poisoned and there tend to be legal costs to pay.

*On balance, it certainly seems more sensible to make a careful soil investigation, define the expected conditions as fully as possible and pay the proven extra costs.*

*It is important to note that because the court found that Wigmar did not rely on the soil reports, the judge was spared the task of deciding whether the reports contained errors or omissions. Neither did he say whether the precise terms of the disclaimer protected the owner from such errors or omissions.*

*The disclaimer does not say the owner has no liability if the soil reports are wrong or omit key information, or that the owner has no liability if it has turned over some but not all of the pertinent soil reports it has in hand.*

*It is not hard to see this owner in the same position as the owner in the Brown & Huston case, which is summarized later in this book.*

*The following case illustrates an application of the Edgeworth decision to a subcontractor claim.*

## J. P. METAL MASTERS INC. v. DAVID MITCHELL CO.

Court of Appeal of British Columbia; March 1998

In 1995, Busby Bridger, an architectural firm in Vancouver, was hired by Bollum's Books Ltd. to act as project architect for redeveloping an old building into a book store. David Mitchell Co. Ltd. was the general contractor and J.P. Metal Masters Inc. was the structural steel subcontractor.

Among other things, the architectural drawings — detail #24 — showed steel brackets intended to support a number of bookshelves. When work started on the project, it was discovered that the bracket system could not be implemented as designed. Busby Bridger redesigned the brackets and revised their spacing but every connection had to be individually engineered on site. Due to these changes, the cost of the work to Metal Masters was much higher than estimated.

Metal Masters sued Mitchell for extras amounting to $75,735 and Mitchell filed a third party notice against Busby Bridger seeking indemnification for any liability it might have to Metal Masters. Busby Bridger applied to the courts to have the action dismissed, but failed. The architect appealed.

Of key interest in the appeal was the issue of buildability and negligent misrepresentation. The contractor claimed that the architect negligently misrepresented that its design of the brackets was buildable when, in fact, it was not.

Busby Bridger conceded that the design of the brackets was not buildable as shown but argued that detail #24 was only a drawing intended to show how the completed construction was to look — that is, it was a design intent only or, as its counsel put it, a concept as opposed to a specification.

However, there was nothing on detail #24 to alert the bidders that it was only a concept and that it was the responsibility of the bidder to ascertain whether it was buildable, and to bid accordingly. Said Justice Hollinrake of the Court of Appeal:

> In my opinion, the evidence can lead only to the inference that as far as Mitchell and Metal Masters were concerned the construction as shown in detail 24 was "buildable" as it appeared on that drawing. That is the misrepresentation the trial judge found and the misrepresentation she inferred the respondents relied on in making their bids. In my opinion, it was clearly open to her to make this finding and to draw that inference. Further, I think the evidence supports her conclusion that the reliance asserted by the respondents was a reasonable one in all the circumstances of this case.

He quoted from the decision of the Supreme Court of Canada in *Edgeworth Construction Ltd. v. N.D. Lea & Associates*. In that case, the general contractor sued the design engineer claiming negligent misrepresentation in the bid documents. Justice McLachlin said:

> One important policy consideration weighs against the engineering firm. If the engineering firm is correct, then contractors bidding on construction contracts will be obliged to do their own engineering. In the typically short period allowed for the filing of tenders — in this case about two weeks — the contractor would be obliged, at the very least, to conduct a thorough professional review of the accuracy of the engineering design and information, work which in this case took over two years. The task would be difficult, if not impossible. Moreover, each tendering contractor would be obliged to hire its own engineers and repeat a process already undertaken by the owner. The result would be that the engineering for the job would be done not just once, by the engineers hired by the owner, but a number of times. This duplication of effort would doubtless be reflected in higher bid prices, and ultimately, a greater cost to the public which ultimately bears the cost of road construction. From an economic point of view, it makes more sense for one engineering firm to do the engineering work, which the contractors in turn are entitled to rely on, absent disclaimers or limitations on the part of the firm. In fact, the short tender period suggests that in reality this is the way the process works; contractors who wish to bid have no choice but to rely on the design and documents prepared by the engineering firm. It is on this basis that they submit their bids and on this basis that the successful bidder enters into the contract....

In the scheme of things, it makes good practical and economic sense to place the responsibility for the adequacy of the design on the shoulders of the designing engineering firm, assuming reasonable reliance and barring disclaimers. The risk of liability to compensate third parties for design error will be reflected in the cost of the engineers' services to the owner inviting tenders. But that is a much better result than requiring the owner to pay not only the engineering firm which it retains, but indirectly, the additional engineers which all tendering parties would otherwise be required to retain.

Busby Bridger argued, however, that Mitchell and Metal Masters did not rely on detail #24 in the drawings put out to bid.

Justice Hollinrake agreed that Mitchell and Metal Masters did not rely on detail #24 as being specifications: those were supplied by the subcontractor in the shop drawings. However, this did not help the architect, because it was irrelevant. Mitchell and Metal Masters relied on the representation that detail #24 showed what construction was to be done and that the work as shown in the detail was buildable. This could be inferred from the fact that they bid on the work.

The court unanimously dismissed the architect's appeal.

---

*For years, it has been a hard sell to persuade a court that a set of bid documents are also the consultant's representation that what is described in the bid documents can actually be built. J. P. Metal Masters appears to have made just such a finding. Part of the court's reluctance — in the past — was that it put too much responsibility on the consultant when, after all, the contractor was supposed to be the expert at building things.*

## WIB CO. CONSTRUCTION LTD. v. SCHOOL DISTRICT NO. 23 (CENTRAL OKANAGAN)

Supreme Court of British Columbia; June 1998

In early 1995, WIB Co. Construction Ltd. was the lowest bidder for the construction of a school for School District No. 23 (Central Okanagan) in British Columbia. However, the District's architect, Fulker Maltby Architects Inc., disqualified WIB's bid and accepted the next lowest. WIB sued the District for breach of contract. It also sought damages from the architect for acting negligently and in bad faith, and for inducing a breach of contract by the District.

One of the main issues at the trial was whether the District was entitled to rely on the advice of its architect that WIB was not capable of performing the construction of the school. The other key issue was whether Fulker Maltby's advice was justified.

Fulker Maltby knew WIB from a previous project, the Extended Care Unit (ECU) of the Kelowna General Hospital, for which Fulker Maltby was the architect and WIB, the general contractor. The project was plagued by serious problems. Fulker Maltby and WIB blamed each other for most of those problems.

In its suit against the District, WIB's position at the trial was that the architect, as a result of those old disputes, was in a conflict of interest, that it was prejudiced and unfair, and that it acted in bad faith when it advised the District not to award the school contract to WIB.

Much of the trial centred on the difficulties encountered on the ECU project and the relationship between Fulker Maltby's project architect, Purrl, and WIB's site superintendent, Mr. Avery. The essence of the architect's complaint was not so much that Avery was incompetent, but that he was required to spend too much time in the office doing work normally done by the general contractor's project manager rather than day-to-day site supervision required by his job as superintendent.

WIB blamed many of the problems on the excessive number of change orders caused by the architect's errors and omissions. Fulker Maltby contended that, given the size of the project, the number of change orders, though large, was not excessive. The main difficulty was that WIB's quotes for the changes were too high.

Justice Lamperson reviewed the law of bidding and the construction industry custom in British Columbia with regard to the award of contracts. He found that the industry custom and practice was to base the decision on two criteria:

- that the bid must meet all the requirements of the bid documents; and
- that the bidder must be qualified to do the work.

He also held that it was an implied term of the bidding contract (Contract A) that the owner must consider all the bids in good faith and apply the criteria established in the call for bids fairly and impartially.

The parties agreed that WIB's bid met all the requirements and that as the lowest bidder, it should have been awarded the contract, provided it was also qualified to do the work.

In this case, however, the call for bids contained the unusual requirement that the bidders must provide the name and résumé of the site superintendent who would be assigned to the project. WIB named Avery. On the advice of Fulker Maltby, the District concluded that WIB was not qualified because Avery could not supervise the project properly. The advice was based largely on the architect's experience with WIB and Avery on the ECU project.

The court then focused on the way Mr. Vance, a partner with Fulker Maltby, evaluated WIB's bid.

Vance was not directly involved with the ECU project, but he talked to the people who were. Among others, he discussed the ECU project with Kelowna Hospital's clerk of works, who was unwilling to recommend WIB. He also consulted the owner's representatives on two other WIB projects and took into account that some of the problems resulted from errors in the design drawings.

Knowing that rejecting a low bid was a serious matter that could result in a lawsuit, Vance called a meeting with representatives of WIB and the School District. The focus of the meeting was the question of supervision on the school project. WIB was unwilling to significantly increase its resources.

Vance eventually recommended that the School District award the contract to the second low bidder, whose bid was only about $7,000 higher than WIB's, and strongly suggested that the District first examine this issue with the Ministry of Education and the District's legal counsel. The District followed his advice and awarded the contract to the second low bidder.

The court found that the District was entitled to place considerable importance on the identity and qualifications of the site superintendent in making its decision, as long as the decision was made fairly and objectively.

WIB argued that Vance should have had an independent architect evaluate whether WIB was a qualified bidder. The court rejected this claim. There was no evidence that Vance had a financial interest in the outcome of his decision or that he was motivated by personal animosity. He did not make his decision lightly.

Furthermore, Avery's assignment to the project was an important but not the only factor influencing Vance's decision. He was critical of Avery, but he would, nonetheless, have accepted WIB if the contractor had agreed to make arrangements for proper supervision on the school project.

He was entitled to take his firm's experiences with WIB and Avery into account. "To do otherwise", said the judge, "would have been unfair to the School District. Surely a consultant is entitled to rely on personal experiences when advising a client".

Even if Vance's decision was faulty, that did not mean it was unfair. It was made in good faith, without an improper motive.

Neither was Vance negligent when checking Avery's qualifications. He chose not to check his references because he put more importance on Avery's performance on the ECU project. "Just because Mr. Vance did not exhaust all avenues of investigation does not mean that he was negligent", decided the judge.

In view of these findings, it was not necessary for the judge to consider whether the architect committed the tort of inducing breach of contract. Nevertheless, he quoted from *Spectra Architectural Group Ltd. v. Eldred Sollows Consulting Ltd.*:

> While there appears to be no prior Canadian authority on the point, American case law clearly establishes that no tort of inducing a breach of contract is committed when a defendant gives advice to a contracting party, under a legally recognized duty to the effect that the contracting party has the right in law to terminate the contract.

The court, therefore, dismissed WIB's action against the School District and Fulker Maltby.

After the reasons for judgment were handed down by the court, Fulker Maltby applied for an award of special costs. The architect argued that WIB's allegation of bad faith consumed extra time and required the testimony of a number of experts regarding the customs and practices within the architectural profession and its relationship with the construction industry.

When an allegation of fraud or bad faith has been made and not proved, a court must consider whether the defendant is entitled to special costs for other reasons also. In the words of Justice McKinnon in *Ip v. Insurance Corp. of British Columbia*:

> An allegation of fraud, wilful misstatements, or other such claims made against a person casts a serious pall over his or her reputation in the community. Very careful consideration must be given to the defendant before making such serious allegations. At the very least, a prima facie case must exist and if it does not then special costs by way of "chastisement" is a reminder to the [plaintiff] to exercise better care in the future.

At the trial of this issue, Justice Lamperson found that WIB's allegation of bad faith was unfounded — but was it so totally unfounded as to justify awarding special costs to Fulker Maltby? The judge did not think it was.

The situation was not so clear when considered from WIB's point of view at the time when it initiated the lawsuit. Thus, the allegation was not arbitrary. Instead of having to pay special costs, WIB should be given a lesser form of rebuke.

The judge also had to consider the increased length and complexity of the trial due to WIB's allegations, especially since WIB persevered with the allegation of bad faith to the end of the trial.

The scale of costs that the courts award to the successful party at trial, said Justice Lamperson, was meant to indemnify the party for roughly 50% of its actual costs. Inflation has over the years reduced that percentage. However, the rules pertaining to costs do not articulate the 50% principle, so the courts have held that the mere fact that the award of costs falls short of that figure is not enough, in general, to trigger an award for increased costs.

In some circumstances, the courts will, nevertheless, consider the difference between the actual costs of the successful party and the cost entitlement under the court rules. The greater the differential, the greater the inference that increased costs may be warranted. This is especially so when some particular or unusual feature of the litigation, generally caused by the losing party, escalates the costs beyond the norm.

Justice Lamperson found such special features in the case before him. Not to award increased costs to Fulker Maltby would, in the circumstances, lead to an unjust result. Accordingly, he decided that the architect should have two-thirds of special costs.

---

*Earlier in this chapter, in connection with the Edgeworth case, we stressed the importance of disclaimers of liability for professionals trying to protect themselves from the consequences of their less-than-perfect drawings and specifications.*

*The following case saw the bid documents include a disclaimer clause. But, before any court has to determine whether a disclaimer clause is effective, the elements of negligence have to be proven (see the Hedley Byrne summary in Chapter 7 of this book). If those elements are not found to exist, the disclaimer doesn't matter.*

*The case also includes a senior judge's comments on the practical results of the ever expanding liability for negligence — on the one hand, the contractor gains the tools to go directly after the negligent consultant that caused its losses. But the cost and delay of the legal actions skyrocket.*

## JJM CONSTRUCTION LTD. v. SANDSPIT HARBOUR SOCIETY

Court of Appeal for British Columbia; November 1999 & January 2000 (suppl.)

The construction of the breakwater for the small craft harbour developed at Sandspit in the Queen Charlotte Islands gave rise to a dispute between the owner (Sandspit Harbour Society), the project engineer (Westmar Consultants Inc.) and the general contractor (JJM Construction Ltd.).

JJM received the tender package from Westmar on August 1, 1995. Bids closed on August 24. About 80,000 cubic metres of rock was needed to build the breakwater and revetments to Westmar's design; the specifications called for many gradations of rock. The proposed contract was based on the standard CCDC 4-unit price form.

The bid documents referred to a geotechnical report prepared for Westmar in 1993 by Thurber Engineering Ltd. There was also a reference to a sampling report done on some of the rock in the local quarries. What was not included in the bid documents was a second report prepared by Thurber in May of 1995. The earlier geotechnical information was non-committal on the availability of suitable local rock. The May 1995 Thurber report was more negative; it included the statement:

> It would be imprudent to assume (suitable material) is available from (either of the local) quarries based on available information.

The bid documents included a disclaimer clause which read, in part, as follows:

> The following documents (listed) are bound into the bid documents but will not form part of the Contract and are not intended as a representation or warranty but are furnished in order that the Bidder may have access to the same information which is available to the Owner. The Owner will not be responsible for any deduction, interpretation or conclusion drawn therefrom by the Bidder....

A letter Westmar sent out with the bid package named two local sources, which we shall call Pit A and Pit B, as "potential rock sources", but the letter contained the following statement:

> Please make your own determination of these and other rock sources that you may wish to consider.

Ultimately, a contract was negotiated. JJM nominated Pit A as the rock supplier. However, Pit A could not meet the requirements, either in the quality or quantity of stone required by the specifications. JJM had to go find the stone in Alaska.

There was a serious dispute between JJM and the Sandspit Harbour Society. The Society terminated JJM's contract, and another contractor finished building the harbour. JJM sued Sandspit and Westmar, claiming that they, in their bid documents, negligently misrepresented to JJM and other bidders that:

- Westmar had conducted sufficient and proper investigation and had obtained reasonable and proper assurances concerning the ability of rock suppliers to supply the rock required by Westmar's design specifications; and
- rock in sufficient quantities and of sufficient quality would be available from the two pits near Sandspit.

The summary addresses only the main points of the claim focusing on negligent misrepresentation.

At trial, the court found that JJM had bid the job, assuming that the required stone was easily obtainable, but had not made a full investigation into the proposed sources. Neither did Westmar assess the site in order to design the breakwater based on the rock that could be obtained from it.

However, Justice Lowry found nothing in the bid documents that represented that Westmar had assumed, or was prepared to assume, the risk of having bidders rely on any supposed assessment it had made. On the contrary, the covering letter clearly stated that bidders were to make their own determination of the rock sources.

Normally, the supply of materials, labour and equipment is the exclusive purview of the contractor. JJM argued that Westmar "designed" the rock for the breakwater — presumably in the way that an engineer designs the aggregate/cement mix for a concrete structure. If that was so, the source of the rock would become a design issue, and Westmar would possibly be liable to JJM on the basis of the decision of the Supreme Court of Canada in *Edgeworth Construction Ltd. v. N.D. Lea & Associates Ltd.*

The trial judge did not accept this argument. Engineers do not design rock. They design the structure and specify the material required to build it. The contractor finds the source or sources from which the material can be obtained to meet the specifications, and at acceptable cost.

JJM further argued that it would be nonsensical for an engineer to design a breakwater without knowing that there was at least one reasonably accessible source of rock available that would meet the specifications.

Generally, responded Justice Lowry, the consultant preparing the bid documents owes a duty to bidders to exercise reasonable care that the information presented reflects with reasonable accuracy the nature of the work so as to enable the contractor to prepare a proper bid. This issue was settled in the 1984 decision of the Ontario High Court in *Cardinal Construction Ltd. v. Brockville (City)*.

The consultant is under no obligation, continued the judge, to determine for the benefit of bidders that the materials required for what is to be constructed are available at a price the bidders might consider acceptable, and there is no implied representation in that regard:

> In my view, a case based on a representation about the availability of *all* the rock to build the breakwater, whether from [Pit A] or elsewhere, does not even get started: there was no such representation.

The trial judge was never put to the exercise of parsing the disclaimer clause because he concluded that JJM had not relied on the geotechnical information in the bid package. Having reviewed a complex fact situation, including a lack of any complaints from JJM until long after the local quarries had failed it, Justice Lowry included the following comments:

I find that [JJM] did not rely on the tender documents with respect to the gradations of rock the undeveloped quarry could produce and that the 1995 Thurber memorandum would not have had any real effect on it entering into the contract at the price his company did had it been included in the tender package.

Having made this finding, Justice Lowry rendered irrelevant some findings of fact which he had previously made concerning the basis of Westmar's estimate for the project and Westmar's motivation for not disclosing the Thurber memorandum:

> The engineer's proposal was nonetheless stated to be premised on their assumption "that a good quality quarry near the harbour site, which could produce competitively priced rock, would be used as the primary source of the material for the breakwater", although a 15% contingency was said to be built into the estimated cost to cover the "uncertainty" associated with the lack of an undeveloped quarry.
>
> ... Mr. Allyn [of Westmar] says that he decided not to include the 1995 Thurber memorandum in the tender package because it was not, in his view, a reliable document ... it was not only left out of the tender documents, but, as indicated, it was not given to the Society with the proposal.
> ... It should be remembered that Westmar was competing with Hayco for the project and the engineers knew both that the Society wanted to utilize local resources and that there were clear cost restrictions on what could be built. Another explanation for the Thurber memorandum not being disclosed, and one that may be difficult to resist, is that Westmar feared that by releasing the geotechnical opinion they would emphasize an uncertainty that would undermine the competitiveness of their cost estimate, predicated largely as it was on the successful development of a local resource that the Society wished to have used. And, having withheld the 1995 memorandum through the proposal and tendering stages of the project, the engineers may have feared the effect if not the embarrassment of disclosing it in January 1996 when it was evident that much of the rock would have to come from offshore.

JJM appealed to the Court of Appeal for British Columbia. Madam Justice Southin based her decision on the following propositions of law:

(a) An owner owes a duty to a builder to take reasonable care in the preparation of bid documents. The bidder is entitled to an accurate and full description of the nature of the work and its factual components. Failure by the owner to disclose that which it knows may be a breach of that duty.
(b) The owner and the engineer both impliedly, if not expressly, warrant in the case of the former, and represent in the case of the latter, that the proposed works are "buildable".

As far as the owner is concerned, added the judge, these propositions are subject to whatever disclaimer clauses are in the contract. She further assumed

— but did not decide — that availability of specified materials is part of buildability and that availability contains within it some aspect of affordability.

The trial judge, in his decision, had held that the words in the bid documents concerning potential sources of stone were nothing more than a statement of a possibility. Justice Southin agreed.

With regard to the issue of buildability, the trial judge apparently rejected the proposition that buildability included the availability of specified materials. Justice Southin felt no need to decide whether the trial judge was right or wrong because the facts of the case did not require such a determination.

Rock which would meet the specification was available but, apparently, not from a local source. Westmar had taken the specifications for the rock from a publication of a federal Public Works Department. "One can fairly assume that government departments do not specify the impossible", commented Justice Southin somewhat optimistically. Furthermore, rock which met the specifications was ultimately obtained but at a substantially greater cost than the contractor had allowed for in its bid.

Justice Southin decided that JJM had failed to demonstrate Westmar's negligent misrepresentation and dismissed JJM's appeal on this branch of the case. JJM also lost on the remaining issues omitted from this summary.

In her supplementary reasons (addressing the issue of costs), Justice Southin had some sharp comments with regard to what she called "a very unfortunate extension of the tort of negligence" in the *Edgeworth* decision:

> Shortly put, my point is that the Legislature ought to enact that architects and engineers owe no duty of care in tort to persons other than their clients, and owners letting contracts for the construction of works owe no duty of care either to the general contractor or the subcontractor. I am speaking only of a duty of care as a foundation for an action of negligent misstatement causing economic loss. To put it another way, the Legislature should override the decision in *Edgeworth Construction Ltd. v. N.D. Lea & Associates Ltd.* ...
>
> It would also be a good thing if the Legislature disposed of the doctrine of concurrent liability in contract and tort.

The amount claimed by JJM was approximately $5,000,000. After a trial of 44 days, JJM failed completely.

Westmar paid legal bills of $734,985; part of that was disbursements. The owner's legal bill was $486,000 for fees and approximately $88,000 for disbursements. The contractor's legal bill was $685,000, possibly not including disbursements.

But it is not only the parties that pay, continued the judge. Much judicial time was expended before the action ever reached trial. Judicial stipends, incidental allowances and ultimately pensions come out of the federal treasury. From the provincial treasury comes the cost of the support staff and the maintenance of the court buildings.

By contrast, Justice Southin estimated that if the only cause of action open to JJM had been based on its contract with Sandspit, the trial might have taken 10 days instead of 44. Westmar, which had no contract with JJM, would not have incurred any legal bills, defending itself against the contractor (although Sandspit might have brought some sort of third-party proceedings against it). The costs for the owner and contractor would have been only a quarter of the actual amounts.

---

*When the trial judge found as a fact that JJM had not relied on any of the geotechnical information in the bid documents to make its bid, Westmar got lucky. Because the contractor took the risk of rock supply without giving it much thought, the engineer's failure to deliver the Thurber memorandum became a non-event.*

*As to the high cost of construction litigation, Madam Justice Southin's frustration is shared, no doubt, by the entire construction industry. Her invitation to the legislature of British Columbia may resonate with the design professionals and their insurers, but her views are out of phase with the Supreme Court of Canada which decided Edgeworth in the first place.*

## SUMMARY

The following points apply to consultants involved in the process of bidding:

- The bid documents must be prepared keeping the average bidder in mind, not the bidder that has special knowledge or experience.
- The engineer/architect that prepares bid documents owes a duty of care to bidders that are known to rely on the information contained in the documents.
- The consultant's duty is to exercise reasonable care that the information presented reflects with reasonable accuracy the nature of the work and, perhaps, that it can be built.
- If the consultant has not verified specific information, the consultant has a duty to inform bidders in clear terms that it does not vouch for the accuracy of the information. This way, each bidder knows that it must investigate for itself.
- If the consultant learns during the bid period that some of the information supplied is incorrect or is incomplete, the consultant has a duty to inform bidders of the true facts.

- The errors in consultant design work, with or without a bid package, may be regarded as negligent misrepresentations; therefore, the consultant is liable to those who (reasonably) rely on the consultant's work product — in relation to Contract B — even in the absence of a contract between them.
- The consultant can attempt to protect itself against such liability by disclaimers of liability and adequate insurance.
- When the bid documents make the contractor responsible to choose the method of construction, the consultant has no obligation to anticipate what method the contractor might select. Also, the consultant has no obligation to warn the contractor of any possible problems associated with that method.

## Chapter 9

# SPECIAL REMEDIES

*The purpose of this chapter is to deal with some particular instances of disputes between bidders and public bodies where the bidder has used a remedy different from your everyday civil action which may award damages.*

*In the remedies discussed in this chapter, the successful party does not receive damages, but the courts use the same principle at work in civil actions to dispense relief.*

### THOMAS C. ASSALY CORPORATION LTD. v. R.
Federal Court of Canada; May 1990

The Goods and Services Tax (GST) was not even law yet when it sparked off litigation. On February 13, 1990, Public Works Canada invited submissions from four companies owning buildings in the Ottawa area for the rental of a minimum of 9,000 sq. metres of office space needed for the new GST department. Each company received detailed bid specification.

The documents indicated that all compliant offers would be analyzed and ranked, based on lowest net present value to the Crown. This provision conformed with the federal government's policy on leasing. However, clause 10.2 of the bid documents also provided that:

> The lessee [the Crown] is not obliged to accept any offer.

A week later, the bids were opened and analyzed. They were all technically compliant and were ranked by net present value as follows:

| | |
|---|---|
| The Glenview Corporation | $7,643,000 |
| Thomas C. Assaly Corporation Ltd. | $8,498,000 |
| Seltzer (Ottawa) Realty Associates | $8,631,000 |
| Sakto Development Corporation | $13,917,000 |

However, it soon became apparent to Public Works officials that the bid specifications were themselves inadequate. For example, in both the Assaly and

Glenview bids, the building had other tenants, which would involve additional security at great cost. The minister, therefore, accepted the Seltzer bid.

Glenview and Assaly immediately went to court. Justice Strayer of the Federal Court ordered Public Works to reconsider its decision. Assaly and Glenview had had "very reasonable grounds" to believe that the lowest price would be the deciding factor once a bidder met the specifications set out in the bid documents. Yet their bids failed because of considerations not specifically set out in the documents. Said Justice Strayer:

> A fair procedure requires that the party whose interests are to be affected by a decision be aware of the issues he must address in order to have a chance of succeeding.

He found that Public Works officials had a "duty of fairness" in their dealings with the bidders which the court could enforce by *certiorari*.

> **Certiorari**. This is a public law remedy to control the proper exercise of governmental powers by judicial review. The court is empowered to order decisions of inferior courts, tribunals and administrative authorities to be brought before it, and the court may quash them if they show an error of law or have certain other defects.

The minister decided to cancel the first bid call and ask the same group of four companies for new bids on the basis of improved specifications. The bid documents also provided that offers would be evaluated not strictly on the net present value but also on considerations as to what the Crown perceived to be the "best value to it".

The new bids were all compliant, but this time, Assaly was rated lowest at $6,813,188 and was awarded the contract.

In no time the minister was back in court. Initially, Assaly, Seltzer and Glenview all brought an application seeking *certiorari* to quash the minister's decision to cancel the original bid call. Glenview also sought an order of *mandamus* to award the lease under the original bid call. Assaly, not unexpectedly, discontinued its application, but Seltzer and Glenview persisted.

> **Mandamus**. A Latin expression meaning "we command". This is a court order instructing a tribunal, public official, crown corporation, etc. to perform a specified public duty.

The new dispute was argued in front of Justice Denault of the Federal Court. He found that Glenview had not established any entitlement to *certiorari*. It failed to establish evidence of any regulation or government policy which would fetter the government's contractual freedom to such an extent as to require it to accept a bid it found unacceptable.

Such fetters, if any existed, would call into question the validity of government contracts, since the freedom of contract, that is, the ability of each party to freely accept or reject the terms of an offer or counteroffer, is the basis of every contractual agreement.

Glenview argued that the minister was obliged under the terms of his proposal to accept the lowest bid which met all the stated criteria. This interpretation would greatly curtail the government's freedom of contract, said the judge. It would, in essence, make clause 10.2 (the privilege clause) ineffective.

Moreover, even if the interpretation were correct, Glenview's proper remedy would be to sue the minister for breach of an undertaking made in the bid proposals. It did not constitute grounds for *certiorari* unless Glenview could prove some sort of procedural unfairness.

Glenview's case was different from the one that was previously decided by Justice Strayer where he ordered Public Works to reconsider its contract award. In that case, the unfairness resulted from the fact that the minister had used undeclared criteria as the basis of the decision to award the contract to one bidder while refusing it to others.

When all the original bidders were invited to resubmit based on clearer specifications, there was no longer any ground for alleging procedural unfairness. Justice Denault was satisfied that the minister had based the decisions on reasonable and relevant considerations and had had no other feasible options. Inviting new bids was the fairest thing to do.

In its application Seltzer suggested that it was unfair to proceed to re-tendering once all the bids under the first bid call had become public knowledge. The secrecy fundamental to bidding had disappeared, and, as a result, argued Seltzer, the bidders were no longer bidding "on a level playing field".

The court did not accept this argument. The information was known to all, so the playing field was indeed level. Moreover, it was equally plausible that this knowledge was as much to the bidders' advantage as it was to their disadvantage.

There still remained Glenview's motion for *mandamus* that the contract be awarded to it on the basis of the first bid call. The court refused the motion.

There is no statutory duty on the minister to accept any offer. Moreover, *mandamus* cannot be used to compel the minister to act in a certain manner: it is a judicial tool that can be used to make an official perform his or her duty, but not to dictate the *result* of his or her action.

In dismissing the applications with costs, Justice Denault concluded:

The reversals of fortune of the various parties to these motions have amply demonstrated that unfairness of result, like beauty, is a matter of perception which varies greatly according to one's vantage point. It is with good reason that Courts restrict the granting of extraordinary remedies to matters of procedural unfairness.

---

*In the following case, the court decided that a public owner should receive the same treatment as anybody else.*

## PETER KIEWIT SONS CO. v. RICHMOND (CITY)

Supreme Court of British Columbia; July 1992

Over a hundred years ago, in the 1895 case *Haggerty v. City of Victoria*, the Supreme Court of British Columbia held that, generally, courts should not interfere with municipal corporations in their internal policy and administrative government unless they overstep their powers or commit something obviously unlawful.

In that case, Haggerty sought a court declaration that a contract let by the City of Victoria to a higher bidder was unreasonable, improper and unlawful. The court decided it had no right to interfere:

> The discretionary powers of the Council are not subject to judicial control, except where the power is exceeded or fraud imputed and shown or there is a manifest invasion of private rights.

These words helped the same court make up its mind in a similar case in 1992. The action arose from a call for bids issued by the City of Richmond in January 1992 for the construction of a bridge over the Fraser River. Eight firms responded. Peter Kiewit Sons Co. and Dilcon Constructors Ltd. were among them.

The Instructions to Bidders were carefully drafted to give the owner the maximum freedom of action. The instructions included the following privilege clause:

> The lowest or any tender will not necessarily be accepted. The Owner reserves the right to accept the tender which it deems most advantageous, and the right to reject any or all tenders, in each case without giving any notice. In no event will the Owner be responsible for the costs of preparation or submission of a tender.
>
> Tenders which contain qualifying conditions or otherwise fail to conform to these Instructions to Tenderers may be disqualified or rejected. The Owner, however, may at its sole discretion elect to retain for consideration tenders which are non-conforming because they do not contain the content or form required by these Instructions to Tenderers or because they have not complied with the process for submission set out herein.

The Owner reserves the right, at its discretion, to negotiate with any Tenderer it believes has the most advantageous tender, or with any other Tenderer or Tenderers concurrently. In no event will the Owner be required to offer any modified terms to any other Tenderer prior to entering into a contract with the successful Tenderer and the Owner shall incur no liability to any other Tenderer as a result of such negotiations or modifications.

Four firms were also invited to propose alternative designs. Both Kiewit and Dilcon submitted base bids and alternative bids.

After the bids closed, Richmond, through its engineering consultant, P.B.K. Engineering Ltd., requested modifications to Kiewit's alternate design. Kiewit agreed to amend the design, which was the lowest structurally equivalent bid and looked similar to the base design.

However, staff and members of council were worried about the proposed changes to prices which occurred after the bidding closed. The municipal engineer, therefore, decided to make a recommendation to city council that did not take the revised bids into account. The council then accepted Dilcon's bid. The council was not aware that certain conditions were attached to Dilcon's bid, but it knew that Kiewit's alternate design was the lowest structurally similar bid.

Kiewit applied for judicial review of the decision to award the contract to Dilcon and for an order in the nature of *certiorari* to quash the decision and force Richmond to re-evaluate and consider Kiewit's modified bid.

The issues before the court were as follows:

- Was it right for a court to review Richmond's decision?
- If such judicial review was available, was Richmond's process of awarding the contract fair?

Kiewit relied on *Ron Engineering*. The contractor argued that, as its bid was in accordance with the bid documents, there was a bidding contract (Contract A) between it and the owner.

Kiewit also cited the judgment of Justice Strayer in *Thomas C. Assaly Corp. v. R*. In that case, Assaly was low bidder in a bidding process that had indicated the contract would be awarded to the lowest bidder.

Assaly failed to get the contract and applied to the Federal Court for an order in the nature of *certiorari* to quash the decision rejecting its bid and an order in the nature of *mandamus* to compel the owner to award the contract.

Having reviewed the evidence in the *Assaly* case, Justice Strayer said:

> It appears to me that, in principle, certiorari is a possible remedy in such a case. The Minister and his officials, in rejecting or accepting tenders, performed administrative functions which are ultimately authorized by regulations which are, in turn, authorized by statute. These are not broad policy functions involving unlimited discretion....

> There is therefore attached a duty of fairness which courts can enforce by certiorari, a public law remedy to control the proper exercise of governmental powers....
>
> It is also clear that certiorari is a discretionary remedy, and a court may refuse to grant it if adequate alternative remedies exist. I have concluded, however, that a denial of fairness has been made out, and that there is no adequate alternative remedy to correct this administrative error.

The minister cancelled the first tender call and started again with a new one. He did not challenge the idea that *certiorari* should be available to review administrative decisions taken during the bidding process.

Looking at Kiewit's application, the court had to decide whether a public body is in any different position from a private individual or corporation who bids on a construction project. All of the *private* law remedies arising from contract are available in both situations. Why should the public body, asked Justice Vickers, be faced with additional hurdles in the nature of *public* law remedies — such as *certiorari* — and a judicial review?

Justice Vickers agreed with the decision in *Haggerty v. Victoria* quoted above. The public's interest in the integrity of the bidding process should be equally strong for both public or private sector projects. Private law protection afforded by the law of contract should apply equally to both the public and private sector. If the private sector is not subject to *certiorari* in the circumstances, neither should the public sector be.

The complaint raised by someone who has not been awarded a contract does not allege an abuse of statutory powers. It alleges a cause of action in contract, which has a perfectly satisfactory private law remedy in the form of damages.

For example, in *Chinook Aggregates Ltd. v. Abbotsford*, the municipality failed to award a contract in accordance with the bid documents. The British Columbia Court of Appeal upheld a trial judgment awarding damages equal to the amount of the contractor's expenses in submitting a bid. Similarly, applying ordinary principles of contract law, a similar decision was reached in *Best Cleaners & Contractors Ltd. v. R.*

In *St. Lawrence Cement Inc. v. R.*, the contractor applied for *certiorari* to quash a Ministry of Transportation decision which awarded a construction contract to another bidder after having rejected St. Lawrence's bid as improper. The court said:

> The [contractor] now asks this court to intervene in a business decision which would result in another bidder being treated unfairly. Certiorari is available to preserve order in the legal system by preventing both excesses and abuse of power and to make government operate properly in the public interest, but not to protect private rights.
>
> This is a commercial contract which in no way affects the public interest. I do not consider certiorari appropriate under the circumstances.

Justice Vickers shared this opinion. He decided that as a matter of public policy, it would be inappropriate to allow both a public law and a private law remedy in situations involving government contracts where no particular procedure is prescribed by statute or regulations. In these types of commercial cases, he said, the ordinary rules of private law should apply; judicial review is not appropriate.

Was the process followed by Richmond fair? The court, having refused judicial review, did not have to consider this issue. However, the judge made a brief comment: "... the instructions to tenderers are clear, and I would not have been prepared to set aside the decision had I reached an opposite conclusion on the first issue." He rejected Kiewit's petition, with costs.

## CEGECO CONSTRUCTION LTÉE v. OUIMET

Federal Court of Canada; November 1991

Is an owner obliged, because of procedural fairness in the tendering process, to inform the lowest bidder as to why it was not awarded the contract — and then allow this lowest bidder an opportunity to reply?

On May 25, 1991, Public Works Canada issued a bid call for renovations to a government building in Montreal, estimated at over $10 million. The notice of bid contained the privilege clause:

> The lowest or any tender not necessarily accepted.

After the bids were opened, it was found that the bid made by Cegeco Construction Ltée was the lowest. The person in charge of the project, Gilles Ouimet, began to verify all the information contained in the bids and information he had subsequently obtained from the bidders.

It was important to ensure that the bidders were competent to execute the project within the budget and on time, and that they had a record of similar completed projects in the past.

Ouimet requested such information from Cegeco, but the answers were unsatisfactory. Furthermore, Ouimet received a report from the consultants on the project, Boutros Pratte Architects, and their evaluation of the contractor was not favourable. Ouimet's assessment was supported by a second evaluation prepared by another employee of Public Works. As a result, Ouimet did not recommend that Cegeco be awarded the contract. He informed the president of Cegeco that the assessment of his company was unfavourable, and he gave him a copy of the evaluations.

When the Minister of Public Works awarded the contract to the second lowest bidder, Cegeco requested the Federal Court to issue a writ of *certiorari* to set aside this decision.

Cegeco submitted that no decision regarding the awarding of the contract should have been taken before Cegeco had been given sufficient time to reply to the allegations contained in the two assessments. Counsel for the contractor argued that Public Works had a duty to act fairly, and since it had not given the contractor a chance to respond, it had failed in its duty.

Counsel cited the case of *Thomas C. Assaly Corporation Ltd. v. R.* In that case, the minister did not award the lease contract to the lowest bidder. While the call for bids made the bidders believe that the decision would be made on price only, the decision was finally based on certain inadequacies of the rental space.

In the *Assaly* case, the court decided that in the process of awarding the contract there was a lack of "procedural fairness". The courts are entitled to correct such unfairness by issuing a writ of *certiorari*. The judge put it as follows:

> The Minister and his officials, in rejecting or accepting tenders, performed administrative functions which are ultimately authorized by regulations which are in turn authorized by statute. These are not broad policy functions involving unlimited discretion. Decisions with respect to the acceptance or rejection of bids directly affect the interests of persons invited to bid who have undertaken the trouble and expense to tender and to hold their property available until a decision is made. There is therefore attached a duty of fairness which courts can enforce by certiorari.
>
> ... If this were exclusively a matter of unfairness of result, I am not sure that it would be an appropriate case for judicial intervention. But it appears to be, on the evidence before me, a matter of unfairness of procedure. A fair procedure requires that the party whose interests are to be affected by a decision be aware of the issues he must address in order to have a chance of succeeding.

Counsel for Cegeco suggested that the general principle to be followed by the minister is that the lowest bidder should receive the contract, and only for special reasons should the contract be awarded to someone else. This would protect both the minister and the lowest bidder.

Furthermore, there exists in the construction industry an expectation that, except for special reasons, the lowest bidder should be awarded the contract. If special reasons exist for not awarding the contract to the lowest bidder, then procedural fairness demands that the bidder be forewarned and given the opportunity to reply.

Counsel pointed out that Cegeco could give evidence that some of the allegations in the assessments were incorrect, but Cegeco was not given the opportunity to do so. It was, therefore, denied procedural fairness, and thus, the award decision should be quashed.

Ouimet and the minister admitted that in principle, they had the duty to act fairly. They did, in fact, act fairly in all respects, and there were no flaws in the procedure they followed. The court agreed. The representative of the minister, in this case Ouimet, concluded from an objective analysis of the information that the lowest bidder would not be able to fulfill the contract obligation.

Surely, said the judge, in such a situation the representative can decide, without having acted unfairly, which contractor to award the contract to if the representative does not once again go back to the lowest bidder for still more information. Evidence clearly indicated that Ouimet had made many requests for information, and that what the contractor provided was "most inadequate".

Justice Teitelbaum said:

> ... I am satisfied from all the evidence that [Cegeco] was given full opportunity to provide the information requested, that the information provided by [Cegeco] was carefully and impartially analyzed and a decision was made based on the analysis.

It does not make practical sense, concluded the judge, to oblige the owner to go back to the lowest bidder every time the owner decides not to accept the lowest bid, to supply the lowest bidder with the reasons for the decision and allow a reply, and then to once again consider both the bid and the bidder's reply.

---

*Finally, a case that proves the point we were trying to make in the introduction to this report: it is all very well to know the rules but **"the game ain't over till the fat lady sings"**.*

*In this case, the judge ruled that when bidding, "only one contract is formed, not two". So, what do you do if the lady (or gentleman) sings a different tune than the lead singers and the chorus?*

## ELLIS-DON CONSTRUCTION LTD. v. CANADA (MINISTER OF PUBLIC WORKS)

Federal Court of Canada; April 1992

In December 1991, the Minister of Public Works called for bids to construct the space agency building in Saint-Hubert, Quebec. Six companies were invited to submit lump-sum bids; the price for the work was to include all taxes except the GST.

Two days after the minister issued the bid documents, the province of Quebec adopted a new sales tax structure incorporating GST. A number of contractors wrote to the minister to get additional information on how the new tax would affect the cost of the project.

The minister issued an addendum, which included amended bid documents, and it postponed the closing of the bids from January 31 to March 12, 1991.

At the opening of the bids on March 12, Ellis-Don was the low bidder, but it became apparent that each of the contractors had calculated the tax differently. Only after the closing of the bids was the minister informed that the Government of Canada would be exempt from the new tax which was to come into effect on July 1, 1992.

As a result of this confusion, the minister decided to reject all the bids, as the minister had a right to do under the Instructions to Bidders, and to call for new bids which would leave the tax out entirely. All six bidders were informed of this decision and invited to participate in the new call. On April 3, Ellis-Don filed a claim with the Federal Court asking for an order:

- to cancel the minister's new call for bids;
- to reinstate the original call; and
- to award the contract to Ellis-Don, pursuant to the original call.

Justice Teitelbaum of the Federal Court felt obliged to decide the matter quickly because the bids made for the first call would become void on April 26. He noted that the first call for bids stated that the minister would not necessarily accept the lowest or any of the bids.

Ellis-Don submitted that the minister was bound by Contract A to accept or refuse the bids it received. If the minister decided not to accept any of the bids, it had to have a valid reason for that decision, such as the cost of the project being too high or the project no longer being needed. The contractor argued that the confusion regarding the new tax was not a valid reason since, according to Ellis-Don, no such confusion existed.

Counsel for Ellis-Don submitted a number of cases to indicate what the law was regarding competitive bidding. Justice Teitelbaum did not review any of the cases. He found them inapplicable.

He was satisfied that after the bids were opened, they were each carefully examined. If the minister came to the conclusion that there was confusion amongst most of the bidders with regard to a serious issue — such as the new tax — that was enough for the minister to refuse all bids and ask for new ones. The minister has an obligation to act fairly, said the judge. This means that the minister must ensure that the information it gives is understood by the bidders.

In *Glenview Corporation v. R.*, Glenview attempted to put limits on the minister's privilege clause. Justice Denault stated:

> [Glenview] has failed to establish evidence of any regulation or government policy which would fetter the government's contractual freedom to such an extent as to require it to accept a tender it found unacceptable. Indeed, such fetters, if they ex-

isted, would call into question the validity of government contracts since freedom to contract, that is, the ability of each party to a contract to freely accept or reject the terms of an offer or counteroffer, is the basis of every contractual agreement.

Justice Teitelbaum agreed. He found that the minister had the right, based on the privilege clause, to refuse each of the bids and issue a new call.

Ellis-Don did not get very far with its assertion that it had a bidding contract (Contract A) with the minister.

"With respect, I do not agree", said Justice Teitelbaum. "What is put out by the [Minister] is an *Invitation to Tender* and it is for the [Minister] to determine if he wishes to enter into a contract. It is an Invitation to Tender to sign a contract. Only one contract is formed, not two as [the contractor's] counsel submits."

Although the case was never appealed, it is a safe bet that the judge's finding that there was, in effect, no Contract A would not be applied in today's environment.

## DONOHUE RECYCLING INC. (C.O.B. ABITIBI CONSOLIDATED RECYCLING DIVISION) v. TORONTO (CITY)

Ontario Superior Court of Justice; July 2002

In 2002, Donohue Recycling Inc. responded to a Request for Proposals from the City of Toronto. However, the City advised Donohue by letter that it could not consider its proposal:

> Your submission for Request for Proposal ... has been declared informal as your proposal did not include the required Bid Security as stated in section 4.10.1 'Letter of Credit' which was mandatory to be included at the time your document was submitted.

Donohue asked the Ontario Superior Court for an *injunction* or *mandatory order* to prevent the City from evaluating other proposals and/or ranking such proposals without having considered Donohue's proposal on its merits. Donohue argued that the RFP was ambiguous, and that Donohue had submitted all of the documentation required by the RFP.

Justice Wright observed that the letter from the City advising Donohue that its proposal could not be considered referred to the Bid Security required by section 4.10.1 — but the words "Bid Security" did not appear in that section. Instead, it stated:

> The successful Respondent(s) shall provide a Letter of Credit (or certified cheque) upon execution of the contract.

This requirement clearly was not part of the proposal process but referred to the "successful" respondent providing security "upon execution of the contract." At the end of section 4.10.1, the following statement appeared:

> Respondents must submit an agreement from a Canadian Chartered Bank listed in schedule A or B to the Bank Act (Canada) to provide this irrevocable Letter of Credit, as outlined in Appendix A of this RFP document.

This statement did not say when an agreement from a bank had to be submitted. It simply referred the respondents to Appendix A.

In section 5.13 Submission Format under Section 1 - Declaration Form and Documents, was the following requirement:

> The following items must be provided as a minimum:
>
> (e) Agreement to Provide Irrevocable Letter of Credit – Appendix A.

However, nowhere in Appendix A was there a mention of an agreement to provide an irrevocable letter of credit. This Appendix referred to Bonds and Bonding Requirements, and stated:

> Every quotation must be accompanied by the following security documentation:
>
> (A) Bid Bond
> (B) Agreement to Bond

No mention was made in Appendix A of an agreement from a bank to provide an irrevocable letter of credit. The only reference to a letter of credit in the Appendix was that bidders may submit a letter of credit in place of an Agreement to Bond.

Donohue complied fully with the requirements of Appendix A.

At the end of Appendix A the following provision appeared in bold type:

> Any bid received that does not satisfy the requirements of the Bid Bond and the Agreement to Bond will be declared informal and not be considered.

Justice Wright noted that Appendix A did not say that any bid received which did *not* provide an agreement from a bank would be declared informal, and not be considered but that is exactly what the City had done. It declared Donohue's proposal informal because the bidder had failed to provide an agreement from a bank to provide an irrevocable letter of credit before the closing of proposals on May 27. After that date, Donohue did provide such an agreement.

The court found that the wording of the RFP was ambiguous and, as a result, it had to be interpreted in favour of Donohue and against the drafters of the document. It would have been a very simple matter to have amended Appendix A to add:

(C) Agreement from a bank to provide an irrevocable letter of credit.

There was no such provision.
The court granted the relief requested by Donohue, plus costs.

In *Donohue*, Mr. Justice Blenus Wright granted injunctive relief in a procurement case — an unusual step. His decision flowed from four observations. But, Justice Wright did not state the authority upon which his decision was grounded. Justice Wright's reasons are open to different interpretations.

A call to counsel revealed that Mr. Justice Wright presided over an application under Rule 14.05(3) of the Rules of Civil Procedure. The Rule permits the court to determine, on an application, "...rights that depend on the interpretation of a deed, will, contract or other instrument...or a municipal by-law or resolution." Available relief includes an injunction and/or a mandatory order.

So, Rule 14.05 was the hook upon which Mr. Justice Wright hung his hat. But, even that hook requires something which creates enforceable rights. No deed or will is in the cards. There is no mention of a municipal by-law or resolution. A contract — perchance? The process was called an RFP. So, where is the contract?

Contract A? Of course!

RFPs share something with letters of intent — being misunderstood. Many assume that dubbing a procurement process "RFP" immunizes it from the *Ron Engineering* Contract A. The RFP label must be ignored and the substance analyzed before concluding whether the process creates a Contract A, or not.

A word about public law remedies. When running a procurement process, municipal officials are performing administrative functions. The municipality is moving toward a private law instrument — a contract. At the same time, the municipality may be subject to a public law remedy respecting the way it behaves.

In the past, courts have been asked to apply public law remedies in procurement cases — remedies involving judicial review of the exercise of governmental power. One such remedy is to overturn or quash a government decision. Another is to order a government to do something it has refused to do.

In *Thomas C. Assaly Corporation Ltd. v. R.*, the court quashed a bid process for procedural unfairness, a breach of Contract A. The decision of Mr. Justice Strayer included the following statement:

The Minister and his officials, in rejecting or accepting tenders, performed administrative functions which are ultimately authorized by regulation which are, in turn, authorized by statute. These are not broad policy functions involving unlimited discretion....

There is therefore attached a duty of fairness which courts can enforce by certiorari, ... a public law remedy to control the proper exercise of governmental powers....

Other courts, faced with claims of unfairness in a bid situation, have refused public law remedies because the public interest was not at stake.

In *Donohue*, we must conclude, by inference, that the City's RFP process gave rise to Contract A, following *Ron Engineering*. The City's rejection of Donohue's proposal, on grounds not found in Contract A, was, among other things, unfair. Being unfair is a breach of Contract A.

If Contract A is truly a contract, then classic contract remedies, including injunctive relief, are available.

Donohue's relief was procedural only — the City was obliged to consider Donahue's RFP (*i.e.*, perform Contract A). Donohue would recover no damages, and had no guarantee of Contract B, the recycling contract it wanted.

## SUMMARY

There are two special remedies when dealing with the public sector:

- Public sector employees have a duty of procedural fairness in their dealings with the bidders, and the courts can enforce this duty by a writ of *certiorari*.
- The courts can use a writ of *mandamus* to compel a public official or authority to perform a specified public duty.
- *Mandamus* can be used to compel the official to perform his or her duty, but not to dictate the result of the official's action.
- It is good to know the rules of the bidding game (or any other game), but you can never be sure that the referee will play by the rules.
- Since these remedies deliver neither Contract B nor compensatory damages, wronged bidders should look elsewhere for relief.
- There are also other not so special remedies available which do not result in an award of damages. One of them is an injunction or mandatory order. Both look like the "special" remedies described in *Assaly*. But their authority comes from the authority of the court to, among other things, preserve the status quo through an injunction or direct a party to perform a contract obligation as happened in *Donohue*. The bedrock for the court's intervention is the existence of a relationship between the parties bestowing rights and obligations — like Contract A. After all it is a contract — really.

# Chapter 10

# TIME OF BID SUBMITTAL

*There is no doubt in anybody's mind as to what should be done with a late bid: it must be returned to the bidder unopened. Thus, a bidder that is a minute late may see its efforts go down the drain. That is why the following issue might be crucially important, however trivial it may seem to outsiders.*

*The issue is the one-minute interval immediately after the deadline for bid submittal. If the deadline is, say, at 3:00 p.m., and the bidder submits its bid several seconds later but before the clock shows 3:01 p.m., is the bid late?*

*Two courts arrived at two opposite conclusions.*

### SMITH BROS. & WILSON (B.C.) LTD. v. B.C. HYDRO

Supreme Court of British Columbia; February 1997

The *Smith Bros.* case deserves notice both for what it decided and for what it avoided.

Smith Bros. and four other contractors submitted bids to B.C. Hydro for a construction project. As is common in bids, the bid documents stipulated that bids had to be submitted by a particular time, in this case at 11:00 a.m. at B.C. Hydro's office.

As sometimes happens, Smith Bros. was tight on its timing and submitted its bid at 11:01 a.m. according to the B.C. Hydro clock. The bid clock recorded only hours and minutes, and not seconds.

Initially, B.C. Hydro rejected the bid for being late. A day later, B.C. Hydro reversed itself on the basis that its clock was a little fast and that it had the practice of accepting bids that arrived within the 60-second period following 11:00 a.m. A few days later, B.C. Hydro reverted to its original position that Smith Bros.' bid was late.

During the period in which B.C. Hydro was making up its mind, the second low bidder, Kingston Construction Ltd., became involved. Its lawyer wrote to B.C. Hydro and complained about the Smith Bros.' late bid, as Kingston was now low and likely to get the work.

Smith Bros. brought an action against B.C. Hydro and Kingston, seeking a declaration that Contract A came into existence with the submission of its bid. Smith Bros. also sought an order for *specific performance* requiring B.C. Hydro to treat its bid fairly as it should the other four bids which were submitted.

> **Specific Performance**. A court order to a person under a contract to fulfil his or her obligations.

The presiding judge had to determine the following:

- Was 11:00 a.m. according to the time shown on the B.C. Hydro clock, or the exact time as shown by an accurate clock?
- What time did Smith Bros. actually deliver its bid?
- Was the time limit exactly 11:00 a.m., or did it extend through the next 59 seconds?
- If Smith Bros.' bid was on time, would this give the contractor an enforceable right to have its bid considered with the others?

First, the judge held that 11:00 a.m. meant accurate time and not something that might have appeared on the owner's clock. The plain meaning of the bid documents suggested this as a common sense decision.

On the basis of expert evidence at the trial, the judge concluded that the B.C. Hydro clock was 60 seconds fast and that the Smith Bros.' bid was actually received at some point in the 60-second period between 11:00 a.m. and 11:01 a.m.

As to whether or not the Smith Bros.' bid was "on time", the judge concluded that it was not. Although B.C. Hydro sometimes allowed a bid to be received in the 59-second period after 11:00 a.m., this was not written into the bid documents. The clear provisions of the bid documents must prevail. Smith Bros. was late.

The judge then considered whether Smith Bros. could shoehorn its bid into consideration by B.C. Hydro. On this point, the judge referred to *Ron Engineering*, *Best Cleaners*, *Chinook Aggregates* and *Vachon Construction* to establish, collectively, the proposition that a validly submitted bid gives rise to Contract A.

Contract A would oblige B.C. Hydro, or any owner, to consider the bids received fairly. Finding that there was no difference between amending an invalid bid after the close of bids and accepting a late bid, the judge determined that B.C. Hydro would breach its duty of fairness to the other bidders if it considered the late bid. Put another way, Contract A for Smith Bros. never came into being. There was no contractual foundation upon which specific performance could be ordered.

At the time that the application came to court, the irrevocable period for all bids was still open, and B.C. Hydro had not let a contract. Had Smith Bros. wished to appeal the decision about Contract A, specific performance would have disappeared. Once B.C. Hydro awarded the contract, the only relief left to Smith Bros. was damages.

---

*When bids go awry, litigation usually extends long after the job itself has been completed. Often, the cost of this litigation approaches the amount at stake.*

*Counsel in this case have done their client a major favour. Using the Rules of Court, and co-operation among the parties, the case came before a judge before the contract was awarded.*

*Although Smith Bros. might have been unhappy with the decision, there was some finality brought to the situation; the owner was finally in a position to proceed with evaluating the four bids that were properly before it. In the long term, Smith Bros. did not lose because it still had litigation open to it to recover damages if a higher court could be persuaded that its bid had been on time and Contract A had come into being.*

*The speedy and negative decision (so far as Smith Bros. is concerned) may have provided a powerful disincentive in taking the fight any further. The reasoning was sensible, and a legitimate difference of opinion had been resolved quickly rather than taking on a life of its own.*

*One of the other instructive things that this decision does is recognize that the threshold for Contract A is the timely submission of a bid. If a bid is late, the owner's offer to enter Contract A has expired. The other parties who have established Contract A with the owner have rights against the owner if it breaks the rules of Contract A by letting a latecomer in.*

*The Smith Bros. case settled the question of timing in British Columbia: a second after the appointed hour is too late for the submission of bids. In Ontario, the judge in the following case came to a different conclusion.*

## BRADSCOT (MCL) LTD. v. HAMILTON-WENTWORTH CATHOLIC SCHOOL BOARD

Court of Appeal for Ontario; January 1999

This case dealt with two issues: (a) whether a bid submitted just 30 seconds after the specified time is invalid and should be rejected by the owner; and (b) whether a bid containing an irregularity is void for uncertainty and should be disqualified. The decision of Justice Somers, given in June 1998, was appealed.

The Court of Appeal first addressed the issue of timing.

In 1998, the Hamilton-Wentworth Catholic School Board in Ontario invited five general contractors to submit bids for the construction of a school. The deadline for the submission of bids was changed several times by a sequence of addenda.

Initially, bids were to be submitted "*only until* 15:00 hours (3:00 p.m.)" on April 22, 1998. This was extended to "*only until* 4:00 p.m." on April 30. The deadline was changed again to April 30 "*at* 4:00 p.m." The final addendum stipulated "Friday, May 8, 1998 *at* 1:00 p.m." as the deadline.

The owner's Instructions to Tenderers made it emphatically clear that bids not received by the time stated "WILL NOT be accepted by the owner".

The Instructions reserved to the owner the right to reject any or all bids or to accept any bid. For good measure, the next sentence also provided that the lowest or any bid would not necessarily be accepted.

On May 8, an employee of Bradscot saw the representative of Bondfield Construction Company (1983) Limited place that company's bid on the receptionist's counter after the clock had reached 1:00 p.m. but before it had reached 1:01 p.m. The official clock showed the hours and minutes only, and did not record seconds, but, according to Bradscot, the time was 30 seconds after 1:00 p.m.

In spite of objections, Bondfield's bid was accepted. The School Board took the position that Bondfield's bid was on time because it was submitted before the clock reached 1:01 p.m. The bid at $17,720,000 turned out to be the lowest; Bradscot's bid was just $1,000 more. Bondfield won the contract.

"A bid submitted after the tender deadline is invalid, and an owner that considers a late bid would breach its duty of fairness to the other tenderers", said Justice Laskin. However, he accepted that in deciding to accept Bondfield's bid, the School Board was acting in good faith and trying to comply with its duty of fairness to all bidders.

This was important because Contract A, as defined in the decision of the Supreme Court of Canada in *Ron Engineering*, imposes rights and obligations on bidders and owners that are intended to protect the integrity of the bidding process. One obligation imposed on the owner is a duty of fairness to all bidders.

The wording of the instructions made no difference to Justice Laskin. The words "only until", "at" and "not later than" are all interchangeable. What is needed, continued the judge, is a clear rule as to what they mean.

Unfortunately, the industry did not seem to have a clear rule. The contractors' association considered a bid submitted 30 seconds after the 3:00 p.m. deadline late; the School Board and the architect, as well as the Toronto Bid Depository, were all of the view that any bid received before 3:01 p.m. was on time.

The only other Canadian case which considered this issue was *Smith Bros.* In that case, bids were to be received "until 11:00 a.m." and "not later than 11:00 a.m." In that case, Justice Shaw said:

> In my opinion one cannot read into the quoted words that the time for delivery of tenders will extend past 11:00 a.m. until almost 11:01 a.m.

The trial judge in the *Bradscot* case, Justice Somers, chose not to follow the *Smith Bros.* decision but took the opposite view. Commented the judge:

> In my opinion when it is stated that some deed is to be done "at 2:00 p.m." the time is for that minute and the act is not overdue until the minute hand has moved off the 12 hand to the :01 position.

The judge decided that the timing of Bondfield's bid did not render the bid invalid.

The Court of Appeal saw no reason to interfere. Justice Laskin found the positions of both trial judges reasonable. Both offered the industry a clear rule. Faced with two reasonable interpretations, the Court of Appeal was obliged to defer to the finding of the trial judge. Thus, the decision of Justice Somers now represents the rule that the Ontario industry must follow.

The second issue on appeal was whether Bondfield's bid was invalid because its price was uncertain. The issue arose because Bondfield's bid form was not properly filled out. Bondfield correctly inserted the bid amount in the form in words and figures. However, under the item that followed, which only served to add GST to the net bid amount, there was an obvious calculation error.

Fifteen minutes after the close of bids, the president of Bondfield called the School Board's architect, pointed out the error and confirmed that the net bid price represented the company's official bid. At the architect's request, he confirmed this by fax. The owner then accepted the bid.

Justice Somers said:

> Here, there is no discrepancy between the printed and numerical price. The only problem that arises is in the subsequent calculation paragraph which appears to be superfluous and, on the evidence, was not considered by the owners as being the operative part of the tender.

In the *Ron Engineering* decision, the Supreme Court had this to say regarding a similar clerical error:

> It would be anomalous indeed if the march forward to a construction contract could be halted by a simple omission to insert in the appropriate blank in the contract the number of weeks already specified by the contractor in its tender.

Neither should the clerical error in the Bondfield bid stand in the way of the contracting parties from carrying out their contract to construct the project, concluded Justice Somers, and he dismissed Bradscot's claim.

The Court of Appeal agreed. The error in the calculation of the GST did not disqualify the bid because it did not make the bid price uncertain. That price was set out in the previous paragraph, and the calculation was, if not superfluous, at least subordinate.

## SUMMARY

From a practical standpoint the owner can protect himself or herself against timing problems:

- The Instructions to Tenderers could provide that the time of bid closing be determined according to the owner's clock, whether accurate or not.
- The time of bid closing could be given in terms of hours, minutes and seconds (*i.e.*, 3:00:00).

## Chapter 11

# BID COMPLIANCE AND BID EVALUATION

### INTRODUCTORY NOTE

*Ron Engineering* is to bidding law what Wayne Gretzky was to hockey. When Wayne played, rare was the post-game summary which did not list him in the scoring. In bidding cases, rare are the reasons for judgment free of some reference to *Ron Engineering*. Unlike Gretzky, *Ron Engineering* usually plays for both teams in the same game.

Wayne Gretzky has retired, but *Ron Engineering* has not lost a step in the 20-plus years since it was decided. In fact, *Ron Engineering* has picked up "line mates" which have enhanced its effectiveness. Consider the scoring power of *Ron Engineering* when combined with *M.J.B.*, *Martel* and *Naylor* — three recent decisions from the Supreme Court of Canada.[1]

We are now going to look at bid compliance and bid evaluation, and their challenges, through the prism of *Ron Engineering* and the subsequent decisions.

In this chapter, we take a high-level view of how the industry continues to struggle with bid evaluation, beginning with the fundamental question: Is this bid compliant? Then, if the bid is compliant, and Contract A established, how do we tread the road through Contract A obligations to an award of Contract B — or not?

In Chapter 12, we examine some recurring themes in the determination of bid compliance. Following the decision in *Ron Engineering*, the old labels —

---

[1] Summaries of the decisions mentioned in this and following chapters can be found in earlier chapters of this book. Note that, as before, the name of the case is in italics (e.g. *Martel*), whereas the same word in normal type (Martel) represents the name of one of the parties after which the case is named.

irregular / qualified / conditional / informal — were assigned supporting roles in the determination of compliance / non-compliance / substantial compliance of bids. We also examine some troublesome bid evaluation issues associated with both bidder and owner performance of Contract A.

Chapter 13 is a snapshot of the relief available when the bidding process fails in some way. In the majority of cases, the relief takes the form of damages.

Damage principles are not overly complicated. Proving damages is another matter. It is curious how often a party in bid litigation seems pre-occupied with proving it was wronged, devoting insufficient resources to the proof of its damages. Or, it fails to realize that the available "payoff" comes at too high a price.

Finally, after looking at what happens if Contract A is breached — not a nice picture, really — we examine a largely neglected side of Contract A. As a contract, Contract A can be used as a tool to shape the bid process and control its outcomes.

## COMPLIANCE

To determine whether the parties have crossed the threshold and entered Contract A, two things have to be assessed.

First, did the owner prepare an *offer* (be it called a bid document or something else) for responding contractors to enter Contract A?

Second, if the owner issued an offer to enter Contract A, did the bidder accept that offer by submitting a *compliant* bid?

If both questions are answered "yes," then Contract A exists and the bidder and owner are contractually bound to the terms of the bid documents. For the owner, evaluation of bids comes next.

When an owner plans a procurement process, it needs to decide what kind of process it wants to initiate. Then, the procurement documents need to communicate the owner's intention, whether it be to create Contract A or not. When an owner fails to make up its mind and then act accordingly, it runs the risk of extending an offer to create Contract A when all it wants to do is start talking.

If the underlying decision is to create Contract A obligations in a bidding process, the next step is to determine the amount of discretion or leeway to be created when evaluating a bid and awarding (or not awarding) Contract B.

Most owners draft bid documents intending to create Contract A with compliant bidders. In what follows, we will look at situations where the owner intends to create Contract A, as well as when it does not. We will also look at one instance where the owner says one thing but actually does another.

## Owner Obligations/Owner Control

The Canadian Construction Documents Committee (CCDC) and the Canadian Construction Association (CCA) both have vested interests in the integrity of the bid system.

In 1982, CCDC issued Document 23 entitled *A Guide to Calling Bids and Awarding Contracts* (CCDC 23). As noted in Chapter 12, CCDC23 is currently under revision.

CCA issued its own guide in Document 29 (1995) entitled *A Guide on Standard Contracting and Bidding Procedures* (CCA 29).

In the ideal and pre-*Ron Engineering* world outlined in CCDC 23 and CCA 29, a "formal" bid which was submitted by the time set for closing would require that the contract be awarded to the lowest bidder.

Both CCDC 23 and CCA 29 suggest that, if there is a latent/invisible mistake in a bid which is disclosed before the owner has accepted the bid, the bidder should be freed from its commitment of irrevocability.

The approach to latent/invisible mistake described in CCDC 23 and CCA 29 is a reasonable thing to hope for: let the bidder go. But, *Ron Engineering* changed all that by shifting the risk of an invisible mistake to the bidder where the bid is compliant and includes a promise of irrevocability.

In *Ron Engineering*, the court spent little time on the owner's obligations because, in that case, they were not relevant. But the reasons for judgment included a sentence that gave a preview of what would develop in the wake of the decision. Contract A obligations were *mutual*. The owner's obligations were both "qualified" and "controlled" by what the owner put in the bid documents.

Owners have always tried to control their obligations in bidding situations. When *Ron Engineering* created Contract A, it set those obligations into a contractual framework which meant that a party that failed to meet its obligations could be liable for damages for breach of contract.

Bid documents often contain standard language or import language which has had the benefit of review in the courts. These attempts at control (predictability) are often frustrated by the simple truth that no two fact situations in a bid scenario are exactly alike.

A mismatch between particular fact situations and the case law has led to a proliferation of bid-related litigation. Owner attempts at control frequently fail where capricious facts intervene or where the situation that has arisen was simply not foreseen (see comments concerning "tied bids" in Chapter 12 for an example of the unforeseen).

Very few owners create bid documents which specifically offer Contract A and then tell the bidder how to accept the offer and describe the consideration that will flow once Contract A is entered. If Contract A is treated like the contract it is, that may change. In the meantime, it is reasonable to assume that a compliant bid will give rise to Contract A where the bid documents stipulate a deadline for submission, irrevocability of the offer for Contract B, bid security, an agreement to bond Contract B (if awarded), and Contract B as the prize.

It is open to an owner to attempt to prevent the creation of Contract A by using a disclaimer. Owners determined to approach the bidding process in this fashion may find themselves confronted by a court which applies the "duck principle": if it looks like, sounds like, walks like and otherwise behaves like a duck, then it is a duck.

If the owner's creature works like Contract A, the judge will usually decide it is Contract A, no matter what the owner says. One can imagine a judge deciding that the "sanctity" of the bidding process must override the owner's attempt to avoid Contract A when all the other badges of the Contract A/Contract B model are clearly present.

Bidding being bidding, some owners will issue bid documents which impose Contract A obligations on the bidder but leave the owner with a great deal of flexibility. Examples of such clauses appear later in this Chapter.

## When Does Contract A Not Arise?

Since *Ron Engineering* was decided in 1981, judges in many provincial courts have been uncomfortable with the notion that merely submitting a bid creates a complex contractual relationship. *Ron Engineering* did not stand for the proposition that Contract A arose every time a bid was submitted, although many read it that way.

If *Ron Engineering* needed clarification in that regard, that clarification was received in the *M.J.B.* case. Justice Iacobucci confirmed that Contract A was not automatic, and he included the following statement in his reasons:

> What is important, therefore, is that the submission of a tender in response to an invitation to tender may give rise to contractual obligations, quite apart from the obligations associated with the construction contract to be entered into upon the acceptance of a tender, depending upon whether the parties intend to initiate contractual relations by the submission of a bid.

Whether Contract A comes into existence or not is controlled by the terms of the bid documents. One can imagine terms that could be added to bid documents to exclude the creation of Contract A. But if no contract is created, then neither party owes the other any contractual obligations. If an owner is serious about getting competitive prices on a project the owner wishes to undertake, it may not make business sense to give up the benefits which Contract A provides.

There have been many decisions where courts have held that Contract A did not come into being because the bidder failed to accept the Contract A offer.

Examples of these are *Midwest Management* (non-compliant bid) and *Smith Bros.* (late bid). In both of these cases, Contract A would have come into existence had the bidder complied with the bid documents.

However, there is a shortage of case law where the courts have examined the process put in place by the owner and decided it was not an offer to create Contract A even if the bidders followed it to the letter.

In *Mellco Developments*, the owner instituted a request-for-proposals (RFP) process which included a statement that the process was not a bid call. The subject matter of the RFP was the sale and development of real estate.

The claimant felt wronged by the activities of the owner. It alleged that the RFP created Contract A, and that the owner failed to meet its obligations in the way it ran the process. Because the owner had been so clear as to the nature of the process it was initiating, the Manitoba Court of Appeal held that the process was "merely a request for proposals opening up a process of negotiation." Not all roads lead to Contract A.

Whether by accident or by design, it is always open to an owner to procure goods or services through other means such as open market negotiations.

The owner that wants submissions from interested parties but does not wish to create Contract A, may choose to issue an RFP. Properly drawn, an RFP asks parties for expressions of interest, and sets out the owner's intention to consider those expressions of interest and then to undertake negotiations with one or more parties whose proposal(s) appeal to the owner.

The RFP process will not create contractual obligations but may create obligations of another kind.

For example, if the RFP promises that the owner will select the proponent from those that signify interest and will proceed with the project, it may face misrepresentation claims if it does not proceed, or if it hires someone from outside the RFP process after proponents have spent money making proposals. This result can usually be avoided by disclaimer language which (for example) allows the owner to cancel the process at any time, and which obliges proponents to acknowledge that they are undertaking expenditures entirely at their own risk.

In the last several years in Ontario, there have been numerous large projects billed as public/private partnerships. These and other large and complex projects, which have often been approached on a design/build basis, can aptly fall under the "duck principle" umbrella.

A good example of the "duck principle" in an RFP setting is provided by a recent large project in Toronto. The owner municipality had a very large civil job to do that was also very complicated. Time was pressing, and there was a deadline on the funding which could not be met if the project was done in the traditional way (by preparing drawings and specifications and calling for bids).

The municipality in question initiated a two-part RFP process. This process required qualified proponents to submit two packages at a stipulated closing time (*quack*). One package contained the technical proposal of that proponent (that is, how it would design and execute the job). The second package was the proponent's price for carrying out the work if its design/build solution was selected.

Within the process, the owner had a stipulated period of time within which to review the design/build package against the technical criteria it had established for the project. When that period was over, the owner returned the price package to the proponents whose design/build technical proposal did not pass muster. Then the owner opened the price packages that did.

The financial package for each proponent included its design/build price as a lump sum (*quack*). The price was irrevocable for a period of time and was backed by bid security and agreement to bond (*quack*). The bidder undertook to sign the construction contract if the bidder was awarded the work (*quack*).

Nowhere in the bid documents was the word "bid", "tender" or any of the other nomenclature of the usual bid process. Was this a process which gave rise to Contract A? *Quack*!

Rather than trying to avoid the creation of Contract A, owners seem to be creating bid documents that give rise to a Contract A and deal most of the face cards to the owner.

## Contractors/Subcontractors

It wasn't long after the decision in *Ron Engineering* before Contract A became part of the contractor/subcontractor landscape.

For the next 20 years, Contract A was defined and redefined along a winding road which ended at *Naylor*. What *M.J.B.* and *Martel* did for the privilege clause/fairness, *Naylor* did for the inception and terms of Contract A between contractors and trades.

The situation in *Naylor* involves CCDC2-82 and the use of a bid depository for the mechanical and electrical trades. Naylor was the electrical subcontractor. The principal terms of Contract A are set out in the case summary in Chapter 4. To consider Contract A in a contractor/trade situation, the contractor must be bidding to an owner under bid documents which themselves give rise to Contract A. Without that pre-existing circumstance, the protection afforded the trade under Contract A evaporates.

If Contract A can arise in a contractor/trade setting, the timing is different and, so are the underlying rules.

*Naylor* confirmed that Contract A arose when the trade contractor, Naylor, was carried by the general contractor in its bid to the owner. So, whether a trade price comes through the bid depository or from vigorous pre-closing negotiations, it follows that there is no Contract A between trades and contractors until the contractor picks the trades it intends to carry.

In a bid depository setting like *Naylor* the process of securing trade pricing by general contractors is an orderly and highly structured affair. Trade offers are made in writing and both the trade and the interested general contractors are restrained in their behaviour by the bid depository rules which are mandated by the bid documents. The trades and the general contractor have the same understanding. Each trade is making its offer through the bid depository to perform some part of the project in the event that the trade is carried and the general contractor is successful with the prime bid. If you consider the difference between a rave and a black tie gala, this is the gala.

Outside the bid depository, gathering trade prices before prime bid closing is a rave. Some quotes are in writing; others by phone. Some offers arrive just before the deadline. Many include margin for negotiation. This kind of price gathering, by far the most common in construction bidding, is a vigorous process involving a series of offers and counter-offers where "shopping" prices is part of the dynamic. Contract A is formed between a contractor and a trade when the trade is carried.

If the bid depository makes orderly that which is otherwise fluid, does it matter whether the contractor finds its trade partner at a gala or at a rave? Certainly, the two are totally different parties. What they have in common are the implications of taking your chosen partner home.

However Contract A is formed, its terms develop in the same way from a combination of the prime bid documents and the case law. Fundamental in *Naylor* was that the prime bid documents were known to all the trades. This is a key ingredient in the existence of Contract A and the driver of its terms.

Assuming that the bid documents incorporate CCDC2–1994 or something like it, the terms of Contract A, as defined by *Naylor*, are likely to be:

- Contract A arises when the trade is carried.
- Absent reasonable owner/contractor objection, the contractor must award subcontracts to carried trades if it is awarded the prime contract.
- Upon award, the trade must perform.
- Reasonable objection must arise *after* prime bid closing and be significant commercially or contractually.
- No post-closing bid shopping is allowed.

So, what do we have?

In a depository situation, the rights and obligations of the parties flow from the bid documents, which usually include the bid depository rules and the prime contract. Contract A comes into existence at the moment the trade is carried.

In a non-depository framework, access by the trades to the prime documents is the key to whether Contract A can arise. If it can, then *Naylor* identifies its birth at the moment a trade is carried. After that, Contract A derives its terms from the prime bid documents which will determine whether "the most important term of Contract A" is aboard. As stated in *Naylor*, that term is: "The assurance of a subcontract to the carried subcontractor, subject to reasonable objection."

# EVALUATION

## Privilege Clause

The privilege clause is the most important and perhaps most misunderstood bid document term.

*Ron Engineering* established that Contract A included "the qualified obligations of the owner to accept the lowest tender". The court finished that thought with the words "... and the degree of this obligation is controlled by the terms and conditions established in the call for tenders." This was a direct reference to the typical privilege clause: "Lowest or any tender not necessarily accepted," and all its variations.

In the years since *Ron Engineering*, owners and contractors have wrestled over the "qualified obligations of the owner to accept the lowest tender."

One of the biggest areas of controversy is just how much room the privilege clause gives an owner to award to other than the lowest bidder.

Until 1997, the case law respecting the privilege clause had a geographical flavour to it.

West of Ontario, the prevailing judicial view was that the privilege clause — an express term of Contract A — did not prevail over the implied term that, absent some flaw in the bid or some compelling business reason, the lowest bidder wins.

From Ontario eastward, courts found that the privilege clause was a complete answer and that this express term could not be displaced by an implied term — something read into Contract A.

The key to understanding the workings of the privilege clause is the decision of the Supreme Court of Canada in *M.J.B.*, already mentioned several times in this book. *M.J.B.* was an intriguing case from Alberta. At trial in 1994 (and in the Alberta Court of Appeal in 1997), the case had a distinctly schizophrenic tinge to it: one foot in the east, one in the west.

The privilege clause was held to be a "complete answer" to M.J.B.'s complaint. But both the trial and appeal courts felt that the owner had been "unfair" and directed the owner to pay all bidders their cost of bid preparation. This tension was resolved, sort of, when *M.J.B.* went to the Supreme Court of Canada in 1999. But first, what came before it.

One of the better examples of the western view of the privilege clause is found in the *Chinook Aggregates* case which ultimately went to the British Columbia Court of Appeal in 1989. The Town awarded the contract to the second lowest bidder, which was local, based on a secret local preference policy. This preference was not set out in the bid documents and was not known to the bidders.

In defending itself, the Town quoted the privilege clause and pointed the court to an Ontario decision called *Elgin*. The Town argued, as *Elgin* had found, that the express privilege clause prevailed over an implied term that clashed with it.

Chinook countered that, absent a flaw in its bid or a compelling business reason, the custom of the trade called for the lowest bid to receive the job. The court declined to follow *Elgin* and held that the implied term should prevail because it would have been "unfair" to give a local bidder a secret advantage over others.

The eastern approach to the privilege clause, at least prior to the *Wimpey* decision (now renamed *Tarmac*), is personified in the *Elgin* case.

In *Elgin* and in other Ontario cases which followed it, there was a feature which distinguished those cases from *Chinook* and other western decisions of the same ilk. In the *Elgin* line of cases, the owners involved did not act in a discriminatory or unfair manner (the owner in *Chinook* had). The *Elgin* court found a reason behind the owner's rejection of the lowest bid, and that reason was both well known to bidders, and had business efficacy (a significantly shorter schedule). This was the critical platform upon which the privilege clause was upheld as the complete answer.

The difference in the decisions between eastern and western Canada was not a difference in judicial thinking so much as the east not having presented a fact situation that was unfair enough to undermine the application of the privilege clause.

The east changed its tune (or played it differently) with the *Wimpey* case (now renamed *Tarmac*), which was decided at trial in September 1997. In *Wimpey*, Hamilton Region called for bids on road work. To bid, all bidders had to prequalify. Wimpey and Dufferin were among the pre-qualified contractors who submitted bids.

The bid documents in the *Wimpey* case were standard issue. They included a 60-day irrevocability clause, bid security and the privilege clause. There was nothing in the bid documents outlining the criteria that would be used by the Region in applying the privilege clause.

Wimpey's bid was in the $8.5 million range, $20,000 below Dufferin's. Dufferin, a local contractor, was quick to point out the benefit of its presence to the Region. The Regional Council met twice in camera and awarded the contract to Dufferin without giving reasons. When the case came to trial, the Region testified that there was no reason for rejecting Wimpey.

The *Wimpey* case was the first case where an Ontario court encountered a situation where an owner pushed the privilege clause too far. The inference drawn by the trial judge was that the Region had, in fact, applied a secret local preference policy, just like the owner in *Chinook Aggregates*. The court awarded Wimpey the profit it would have earned had it received the contract and executed it "economically and skilfully." The Court of Appeal for Ontario upheld the trial court decision in 1999.

The decision in *Wimpey* was not the death knell of the privilege clause. Justice Cameron (the trial judge in *Wimpey*) drew east and west together when he included the following statement in his reasons for judgment:

> This finding does not strip the privilege clause of all meaning when used in a contract such as this. The clause, or one like it, is very common in construction contracts.... It permits the owner to reject the low bid in the case of some "force majeure," or if it decides not to proceed with the project because the bids are above budget, or changed circumstances negate the viability of the project or adversely affect the low bidder's qualifications assumed in the pre-qualification standards. It would also permit rejection based on a pre-published policy.

The message in *Wimpey* is that, if an owner is not awarding to the lowest compliant bidder, it should have a defensible commercial reason for doing so.

There are some cases in this book which look like privilege clause cases but are more subtle than that. Examples of such cases are *Best Cleaners* and *Thompson Brothers*.

In *Thompson Brothers*, the City awarded something other than Contract B to the second bidder on the original bid call. The trial judge would have nothing of the City's argument that the privilege clause justified its rejection of Thompson Brothers' bid. The court held that the City was in breach of Contract A, not for misuse of the privilege clause, but because it awarded something other than Contract B. The trial judge's reasons included the following statement (emphasis added):

> Rather than accepting or rejecting one tender or another, or accepting none of those submitted, which it had the right to do, the City had the Plaintiff and Central bid on entirely different work, the Landfill project. When they received those bids, the City coupled them with the tenders submitted with respect to the Lake project

and awarded the contract to Central who had the lowest cumulative price.. *The contract awarded was not responsive to the tender process....* Where there has been such a breach or breaches, it would be unfair to allow the owner to use a privilege clause as a shield when the other contracting party seeks a remedy....

The court refused to read the privilege clause in a way that would allow the City to use the clause to avoid ever being in breach of Contract A. The court also observed that the whole thing was unfair as bid shopping usually is.

Knitting *Chinook, Wimpey* and *Thompson* together, one is left with the proposition that the privilege clause will not excuse a breach of Contract A.

The Supreme Court of Canada had an opportunity in the *M.J.B.* case to draw all the privilege clause threads together. The Alberta Court of Appeal had taken what looked like a pre-*Wimpey* Ontario approach when it concluded that the privilege clause was "a complete answer to MJB's action." At the same time, the court affirmed the trial judge's finding that the owner had been unfair.

If one remembers that *Ron Engineering* was decided on its narrow facts, then it may not be surprising to find that the Supreme Court of Canada took the same approach to the *M.J.B.* decision when it handed down its ruling in April 1999. The privilege clause, said the court, is not a complete answer to M.J.B.'s complaint. Justice Iacobucci's quote, which is included in the case summary, is worth repeating:

> However, the privilege clause is only one term of Contract A and must be read in harmony with the rest of the Tender Documents. To do otherwise would undermine the rest of the Agreement between the parties.... Therefore, I conclude that the privilege clause is compatible with the obligation to accept only a compliant bid..., the privilege clause is incompatible with an obligation to accept only the lowest compliant bid. With respect to this later proposition, the privilege clause must prevail.

Earlier in the judgment for the Court, Mr. Justice Iacobucci had elaborated on what the privilege clause does and does not permit. That earlier statement read, in part:

> Therefore even where, as in this case, almost nothing separates the tenderers except the different prices they submit, the rejection of the lowest bid would not imply that a bidder could be accepted on the basis of some undisclosed criterion. The discretion to accept not necessarily the lowest bid, retained by the owner through the privilege clause, is a discretion to take a more nuanced view of "cost" than the prices quoted in the tenders...it may also be the case that the owner may include other criteria in the tender package that will be weighed in addition to cost. However, needing to consider "cost" in this manner does not require or indicate that there needs to be a discretion to accept a non-compliant bid.

*M.J.B.* dealt with a short and uncomplicated privilege clause like that used in most bid documents. The Court held no more and no less than it needed to to decide that the narrow language of the privilege clause did not excuse this owner where it awarded Contract B to a non-compliant bidder.

Because the award to a non-compliant bidder was a breach of Contract A, the Court required nothing more, and the national anthem was played. The Court must have found it unnecessary to address the issue of an implied term of fairness because the Court did not even mention it. Some may have been disappointed that the Court did not take the opportunity to accord fairness at least indirect approval. But, the industry did not have to wait very long.

## Fairness in Bidding

*Ron Engineering* shifted the risk of a contractor's latent mistake in a bid from the owner to the contractor. Since then, the courts have spent most of their time determining the rights and obligations of both owner and contractor under Contract A. As reflected in this book, one of the stronger threads running through Contract A is the implied duty of fairness which is being read into bid documents which are silent on the point. The appearance of this implied term set up a natural tension with the privilege clause which the courts were obliged by circumstance to resolve.

This book includes case summaries, starting in the early 1980s, which address and establish the implied duty of fairness. The fact that the courts are prepared to find such a term seems in harmony with the findings in other Canadian contract cases that there is a duty between parties to a contract to treat one another fairly and in good faith. So why not in Contract A? Why not, indeed.

In some ways, the desire of the courts to import fairness and good faith dealing into bid situations parallels two legal camps that developed hundreds of years ago. Put simply, the two camps were the Courts of Law and the Courts of Equity.

Historically, these were two separate court facilities administering justice in quite different ways. The Courts of Law applied the strict letter of the contract and were not much interested in giving relief to a party which got the (perhaps unintended) short end of a bargain.

The Courts of Equity, on the other hand, looked to principles of fairness and whether the party seeking relief came with "clean hands." These courts sought ways to balance the very harsh outcomes that sometimes flowed from the strict application of law or contracts with the notion of fairness and intention.

Those two courts and lines of reasoning merged a long time ago, and our Canadian courts now apply both law and equity, depending on the circumstances. Implying terms of fairness where the bidding documents are silent is a way of

CHAPTER 11: BID COMPLIANCE AND BID EVALUATION  **241**

balancing unstated but reasonable expectations against what parties might get if the contract was followed to the letter.

In March 1985, the Federal Court of Appeal gave judgment in the *Best Cleaners* case. In that case, the owner entered after-bid negotiations with the second lowest bidder and then, in a sort of conspiracy, awarded the second bidder a two-year contract that was a four-year contract in disguise. The bid documents had called for bids on a two-year term. In this higher court, the judgment included the implication of a duty on the owner:

> ... to treat all bidders fairly and not to give any of them an unfair advantage over others.

The *Elgin* case (1987) was not, at first look, a fairness case. As reported earlier, the court held that the privilege clause which stated that the owner was not obliged to accept the lowest (compliant) bid was in conflict with the bidder's attempt to introduce an implied term that the lowest bid should always prevail. The headline — the privilege clause is a complete answer — obscured the court's due consideration of the facts of the case.

Although a duty of fairness was not in the forefront, the *Elgin* trial court reviewed the facts and expressed the view that the way the owner acted was neither arbitrary nor unreasonable, which means that the privilege clause was applied fairly. The court seemed satisfied that there was nothing high-handed or improper going on when the court took a decision that would, on the surface, appear different from the decisions that several western courts would take soon after.

In 1989, the British Columbia Court of Appeal affirmed a lower court ruling in *Chinook* and held that the privilege clause did not protect an owner who applied a local preference policy to bid evaluation without disclosing the policy in the bid documents or informing the bidders. Custom of the trade (lowest bid wins) was implied by the court as was the duty of fairness previously found in the *Best Cleaners* case.

While the *Chinook* court refused to follow *Elgin*, it is not hard to see why. In *Elgin*, the court applied the letter of the law once it was satisfied that nothing underhanded had taken place. In *Chinook*, the court refused to apply the letter of the law because it came to the opposite conclusion about the owner's behaviour. The implication of a duty of fairness in *Elgin* would not have changed the outcome in that case, the failure to imply one in *Chinook* clearly would.

In 1991, the Court of Queen's Bench in Saskatchewan handed down a decision in a privilege clause case called *Kencor Holdings*. Kencor, the lowest bidder, was from Alberta. The second bidder was local. Crown officials recommended an award of contract to Kencor. But the minister, by order in council, awarded it to the local contractor. In court, the Crown attempted to invoke the broad powers of the privilege clause and asserted the proposition that what happened was in the best interests of the Province as contemplated by the express terms of the bid documents.

The court awarded Kencor damages after implying a term of fairness and then finding that it was "a blatant case of unfair and unequal treatment". The court was troubled by the notion that the privilege clause might be construed to enable the Crown not to be accountable for its actions.

In June 1996, another case upholding an implied term of fairness was decided by the British Columbia Court of Appeal in *Vachon Construction*. The lowest bidder in *Vachon* submitted a bid which stated its bid price in figures as one number and in words as another. Upon opening the bid, the bid authority could not tell what the bid price was. After asking the bidder to pick a price, the authority amended the bid by crossing off the higher figure. The owner then awarded the (slightly adjusted) lowest bidder the contract, over the complaints of Vachon.

The lower court decided that the discrepancy in the two numbers was an "irregularity" rather than something more sinister. The Court of Appeal went the other way, and held that price was an essential part of the bid. It also reached back to the implied duty of fairness in *Best Cleaners*. In the final analysis, the court awarded Vachon damages on the grounds that it would be unfair to allow the owner to breach its own rules (no amendments to bids after submission) and then use the privilege clause to defend itself.

In the *Martselos Services* case, the Northwest Territories Court of Appeal declined to imply a contractual duty of fairness even though the case had a bit of an odour. The bid involved janitorial services at Arctic College. The bid documents and process were unremarkable of themselves. The case arose because an officer of the lowest bidder was also an employee of the college. This breached certain college conflict of interest guidelines. But, the conflict did not breach any term of the bid documents.

While the trial court upheld the second bidder's claim, the Court of Appeal would have none of it. Neither was the appellate court persuaded that this was a case where there was a secret agenda or differing treatment of the two bids received by the college. As in Ontario, this court satisfied itself that there was nothing unfair in what occurred and then expressly refused to go outside the four corners of the bid documents to imply a duty of fairness in connection with the "odour."

Then came the Supreme Court of Canada's decision in *Martel*. Not a construction case, *Martel* turned its full attention to Contract A and the implied obligation of fairness about which lower courts had reasoned for nearly 20 years.

A summary of *Martel* is included earlier in this book. The importance of the case justifies a rehash. *Martel* was a landlord hoping to retain Her Majesty as a tenant. Unsuccessful party/party negotiation led to a classic bid process.

The bid documents gave Her Majesty wide berth during bid evaluation. In the privilege clause, Her Majesty reserved the right to consider, besides rent, its tenant costs (fit up).

Martel submitted the lowest compliant bid. During bid analysis, Her Majesty added $1,000,000 to Martel's price for "fit up" and a further $60,000 for a security card system. When it analyzed the other bids, Her Majesty added "fit up" to all of them. But only Martel absorbed the costs of the card system. Martel was now second and lost Her Majesty as a tenant.

Martel started a multi-faceted action. One of the facets was an allegation that Her Majesty breached her Contract A obligation to treat Martel fairly.

The Supreme Court of Canada examined the evaluation provisions in the bid documents, including the privilege clause. The Court concluded that an implied obligation of fairness was owed by Her Majesty to Martel under this particular Contract A.

First, the Court considered Martel's claim that the addition of "fit up" was unfair. Since the bid documents allowed it, and Her Majesty added "fit up" to the other bidders — no unfairness.

The Court then considered Martel's allegation that the $60,000 for the card system was unfair. The Court found no reference to the system in the bid documents. It also found that only Martel was singled out for this charge. The Court concluded that Her Majesty had committed two breaches of Contract A. First, Her Majesty added a $60,000 item which was not disclosed in the bid documents — this was unfair. Second, Her Majesty singled out Martel for this additional cost. This was unequal.

At this point, Martel expected relief. Instead, it hit the wall. Although Her Majesty had been unfair, the undoing of the unfairness — the removal of $60,000 from Martel's bid — still left it second.

In approving what lower courts had been saying for years, the Supreme Court of Canada recognized the implied duty of fairness, and more. The Court said so in words worth repeating:

> While the Lease Tender Document affords the Department wide discretion, this discretion must nevertheless be qualified to the extent that all bidders must be treated equally and fairly. Neither the privilege clause nor the other terms of Contract A nullify this duty. As explained above, such an implied contractual duty is necessary to promote and protect the integrity of the tender system. (paragraph 92).

The cases reviewed by the Supreme Court of Canada included all the "usual suspects." Most of them are discussed in this book.

So, now we have it: the law of the land. Fairness is an implied term of Contract A, unless expressly excluded. The privilege clause will not excuse a breach of the duty of fairness.

As well, fairness seems to have two stems. The more obvious is: don't import undisclosed terms into bid documents. The second stem is the issue of equal treatment. Not fully developed, the implication from *Martel* is that had Her Majesty applied "fit up" to Martel, as provided in the bid documents, but failed to impute the same costs to the other bidders, then Her Majesty would breach the implied duty of fairness.

That second message in *Martel* goes to the essential nature of Contract A — a process contract. If an owner contracts with several bidders to run a bidding process, the implied duty of fairness appears to prevent an owner from, for example, applying evaluation criteria unevenly in order to manipulate the outcome. And why not? How is that any different from allowing one bidder to amend its bid after closing while refusing the same treatment to others?

So, the moral of the *Martel* story is:

- Don't invent criteria not disclosed in the bid documents (unfair);
- Apply the disclosed evaluation criteria evenly (be fair).

In a sense, it's all about contractual accountability.

In the seven cases on fairness just discussed, every one of them dealt with government at some level. The party feeling aggrieved took its story to court and sought damages for breach of contract.

Other bidders dealing with government took their complaints in a totally different direction. The different direction involved what are called *prerogative* remedies, which are of limited use to wronged bidders. Several cases involving prerogative remedies are summarized in Chapter 9 of this book (*Assaly, Kiewit, Cegeco, Ellis-Don*). The nature and limitation of prerogative remedies is dealt with in Chapter 13. At this point, it is sufficient to say that although the prerogative remedies are very different, the duty of fairness has surfaced in them just the same.

## Privilege Clause/Discretion Clause

As this book reports, the courts have been implying terms into bid documents to the chagrin of owners. There is greater readiness to award a wronged contractor the profit the contractor would have earned on Contract B if it can be shown that, properly run, the bid evaluation process would have (should have) resulted in the award of Contract B.

This has tempted owners to develop bid documents which require contractors to submit irrevocable bids with bid security but which relax the owner's obligations and increase the availability of owner discretion. In a way, the courts themselves are helping in this effort by establishing and then expanding a "nuanced view of cost" (see *M.J.B.* and *Sound Contracting*).

Here is an example of a "deluxe" owner clause (emphasis added):

1A. The Bidder acknowledges that the Owner shall have the right to reject any, or all, Tenders for any reason, or to accept any Tender which the Owner in its *sole unfettered discretion* deems most advantageous to itself. The lowest, or any, Tender will not necessarily be accepted and the Owner shall have the *unfettered* right to:

    (i)    accept any regular, unbalanced, informal or qualified Tender;
    (ii)   accept a Tender which is not the lowest Tender; and
    (iii)  reject a Tender that is the lowest Tender even if it is the only Tender received.

1B. The Owner reserves the right to consider, during the evaluation of Tenders:

    (i)    information provided in the Tender document itself;
    (ii)   information provided in response to enquiries of credit and industry references set out in the Tender;
    (iii)  information received in response to enquiries made by the Owner of third parties apart from those disclosed in the Tender in relation to the reputation, reliability, experience and capabilities of the Bidder;
    (iv)  the manner in which the Bidder provides services to others;
    (v)   the experience and qualification of the Bidder's senior management, and project management;
    (vi)  the compliance of the Bidder with the Owner's requirements and specifications; and
    (viii) innovative approaches proposed by the Bidder in the Tender.

1C. The Bidder acknowledges that the Owner may rely upon the criteria which the Owner deems relevant, even though such criteria may not have been disclosed to the Bidder. By submitting a Tender, the Bidder acknowledges the Owner's rights under this Section and absolutely waives any right, or cause of action against the Owner and its consultants, by reason of the Owner's failure to accept the Tender submitted by the Bidder, whether such right or cause of action arises in contract, negligence, or otherwise.

This is a deluxe privilege clause, for sure. First, the owner gives itself "unfettered discretion" to retain for evaluation a non-compliant bid. The implications for bidders who are compliant (and who have Contract A with the owner) are that it is not a breach of Contract A should the owner make an award to a non-compliant bidder. With that provision, one of the essential elements of *M.J.B.* is muted.

The clause also recites eight particular elements that the owner may bring to bear during bid evaluation. If these bidders have been pre-qualified, the owner has imported into Contract A the very thing it assessed, outside of a contract setting, during the pre-qualification phase. Maybe a smart move, maybe not.

Finally, the owner creates a licence for itself to import into the bid evaluation process undisclosed criteria (secret agenda?). And the owner finishes off the thought by including a waiver of action by the bidder against the owner.

Is this a fantasy? Not entirely. The privilege clause in *Midwest Management* and the one recently ruled upon in *Graham* include elements of the clause just quoted.

The best kind of contractual discretion that anyone can have is one that is "unfettered" and can be exercised "unreasonably." Both of those characteristics are inconsistent with a duty of fairness.

The ideal position for an owner in a bid document is to bind the bidders to a period of irrevocability, supported by bid security, while giving the owner the freedom to behave the way the owner did in *Wimpey* — but win.

*Martel* stands for the proposition that the implied duty of fairness will exist unless expressly excluded. Will the above privilege/discretion clause exclude that duty? It might. What it might also do is prevent the formation of Contract A, and leave the parties in a pre-*Ron Engineering* landscape where there are no contractual obligations between the parties. So, a bidder may withdraw a bid which includes a latent mistake as long as it communicates the existence of the mistake before the bid is accepted.

The title "discretion clause" appears to have been coined in the *Graham* decision. In the British Columbia Court of Appeal, the court endorsed the ruling of the trial judge that the exercise by the owner of "sole discretion" was subject to objective review by the Court where that discretion was applied to something which had a common meaning.

In *Graham,* the word which the court put under "objective scrutiny" was "material". Contractors might take some comfort from the court's intervention. How the *Graham* decision will play in the face of an "unfettered discretion" remains to be seen. What the court seems to be trying to protect is the integrity of the law. Had the word "material" not been subject to objective scrutiny, the owner's interpretation would have enabled the owner to create Contract A with a bidder which had submitted a bid with an obvious error on its face.

We should all keep an eye on the "discretion clause." Interesting things are happening.

# Chapter 12

# TROUBLESOME BID ISSUES

The bid issues described in this Chapter fall into two general families. The first relates to compliance. The second covers matters of evaluation. Some of these issues can be said to fall into both families, not just one. Even so, we have separated them, with apologies to those who would separate them differently or not at all.

In Chapter 11, we examined the big picture around Contract A. Here, the issues and situations are more specific, identifying what the authors believe to be recurring themes in bid settings.

## COMPLIANCE ISSUES

### On-Time Bid Submission

In Ontario, as a result of the *Bradscot* case, the closing time for bids may be 59 seconds later than everyone thought. The outcome of that case depended on the wording of the bid documents. *Bradscot* may be a "one-off" case but it bears the stamp of approval of the Court of Appeal for Ontario.

In British Columbia, the *Smith Bros.* case held that the stroke of the appointed hour rendered any bids submitted after that stroke late. Which of these approaches the other common law provinces will follow is anybody's guess. The more practical approach is to use *Smith Bros.* — and to change the way you express time.

Various industry associations have taken a firm policy position on the time of bid submission. The most sensible approach is to specify the time of bid receipt down to the second (*i.e.*, 11:00:00).

Some might think that taking the closing time to the second would end the controversy. Wrong. In the *Smith Bros.* case, the controversy also involved whether or not the closing time was dictated by the owner's clock or something more global like Greenwich mean time. In *Smith Bros.*, the clock on the wall was wrong. In British Columbia, the governing time was the actual time, not what an owner's fast or slow clock might say.

It is open to the owner to specify the time in any way it wishes. A prudent owner will spell out the time to the second and identify the clock on which that time is recorded.

## Is This A Valid Offer?

The construction industry used to answer the above question by describing a bid as either "formal" or "informal."

Before *Ron Engineering*, the assignment of the "formal" or "informal" label was reserved for the outcome of the judgment on whether the bid submitted is an offer for the project. Is the bid signed? If required, is it witnessed and sealed? And so on.

There have been bid documents, *Midwest Management* containing one such set, which gave the owner the right to waive non-execution of the bid. There is little evidence that owners have much appetite for a bid which is not an offer for Contract B. What's the point?

One might wonder if an unsigned offer might be a counter-offer. The answer is no. Whether a bid is an offer or a counter-offer, the word "offer" is dominant. If the bid form is not signed, there is no offer of any description.

It is common to hear the labels "formal" and "informal" being applied today to a bid that one of the parties considers to be compliant or non-compliant. When used in that manner, the term is not particularly helpful.

In the context of this book, a bid that is "informal" (*i.e.*, not properly executed) is non-compliant.

## Plans and Specs?

If the process of determining compliance is sequential, then this is the next question in the progression. Does the bid conform to the bid documents? A negative answer to that very general question renders the bid non-compliant, thus meriting rejection.

In *Midwest Management*, the bid was not "plans and specs" because Midwest excluded the cost of dewatering from its lump sum bid although it was called for in the bid documents. Midwest proposed to dewater on a time-and-materials basis. Some would attach the label "informal" to Midwest's bid. In fact, the bid was qualified and therefore non-compliant.

Put in legal terms, the Midwest bid did not represent an acceptance of the owner's offer to create Contract A. The bid was a counter-offer to create a different Contract A with a variation in pricing (part lump sum, part time and materials). The Midwest bid also failed to respond to the owner's invitation to make an offer to carry out Contract B. The offer made in the Midwest bid was for a different Contract B.

The industry and the courts use the term "qualified" to mean two different things. In *Midwest Management*, qualified means flawed. In another context, qualified may refer to a bidder which has survived the owner's pre-qualification process and/or demonstrated that it has the necessary skills, experience and financial strength to be permitted to bid.

If the qualification in Midwest's bid was translated into a dialogue between owner and bidder, it might sound like this:

> *Owner: Give me your best lump-sum price to do precisely what is in the bid documents, including dewatering.*
>
> *Bidder: My lump-sum price is $X,* ***but*** *I will make an additional charge for dewatering on a time and material basis.*

The tip-off in this *Midwest*-centred example (and in most cases where a bid is qualified) is the word "*but*." If the bidder's response to the owner's offer of Contract A is "*yes, but*", the owner is probably looking at a qualified bid.

A bidder also fails to bid "plans and specs" when the appropriate addenda and notifications are not listed in the bid as required. Most of the time, this happens by accident. The result is that the bidder has not bid on the scope of work — "plans and specs" — upon which it was asked to bid. The bid is non-compliant, and Contract A is not formed.

Many bid documents require the bidder to state the planned completion date. Sometimes there is a requirement to submit a construction schedule. Imagine that a bid requires both and a bidder submits neither. In a project with delivery sensitivity, a common occurrence, the finish date/schedule to completion is a major price-related variable in bid evaluation. When a bid lacks this vital information and when the bid documents require it, the bid is not "plans and specs" and is non-compliant.

## Bid Security/Agreement to Bond

Most bid documents require bid security in the form of a bid bond or certified cheque. The same bid documents could carve out room for an owner to waive this requirement should the bidder fail to comply. Experience suggests that such a waiver is rare.

A lack of bid security when specified points to a sloppy bidder or to one without the financial muscle to provide what's required. The industry regards this term as fundamental. So, a bid which fails to include bid security when required is non-compliant.

In the same book but on a different page is the requirement that the bidder provide an agreement to bond. This requirement is in the same family as bid security but is more fundamental. The ability of a bidder to get bonding suggests that it has sufficient resources, or its principals do, to attract the support of the bonding company. As well, the bond itself provides the owner important security against contractor failure.

Few bid documents give the owner the discretion to waive the requirement that an agreement to bond be provided. Failure to provide an agreement to bond represents non-compliance.

## Failure to Nominate Subcontractors

There is tension between the *Megatech* decision and a widely held owner view that a failure to list subcontractors when required is non-compliance. Often, a problem respecting subtrades arises because the contractor, in fulfilling the requirement to list, lists two or more trades where only one is called for, lists its own forces where that is prohibited, or lists both its own forces and a trade.

While the debate continues within the industry, the counsel of prudence is that failure to list (when required) is non-compliance, and the contractor that gets "cute" with multi-trade listing for the same discipline is just as non-compliant as the contractor that lists no trades.

From a business efficacy point of view, any owner has a vested interest in knowing the identity of (at least) the major trades. Mechanical and electrical systems, in particular, are the guts of most building projects. Few owners want to risk those elements of a project with unknown, possibly unreliable or uncreditworthy, subcontractors — even though the contractor is responsible for them.

And then there is fairness. It is a breach of the owner's implied duty of fairness under Contract A to accept the bid of the contractor who has not listed its subcontractors when the bid documents require it. That contractor has committed itself to a bid price. But, at the same time it has the opportunity to shop subtrade prices when all the other bidders have entered Contract A with their subcontractors, thus freezing their subtrade costs.

## Bidder Error in Bid: Latent Mistake Revisited

The cases reviewed earlier in this book describe a road map for the treatment of different kinds of contractor error. *Ron Engineering* and the few cases involving bidder pricing mistake illustrate the unhappy consequences that befall a bidder who makes a mistake that is not "apparent on the face of its bid." The risk of an invisible or latent bid error lies with the bidder. From a legal perspective, if the error is not "apparent" at the time of bid opening, the bid is compliant no matter how loud the bidder pleads for relief.

While the courts have been constant in leaving the bidder with the risk which it assumed, the industry urges a different course of action based more on morality and fair play than on the current state of the law. CCDC-23 (1982) sets out a prescription for mistaken bids. The authors of the document recommend:

> If the bidder informs the bid-calling authority reasonably promptly after bid closing and before the authority communicates acceptance of the bid, that a serious and demonstrable mistake has been made in his bid and requests to withdraw, he should normally be allowed to do so without penalty.

CCDC-23 is under revision. The draft currently in circulation recognizes the impact of *Ron Engineering* on the whole bidding process. When it comes to dealing with invisible mistake, the new CCDC-23 makes the same plea as before: let the unfortunate bidder off the hook.

Where the bidder's error is "apparent" at bid opening, the law of mistake prevents an owner from accepting the mistaken bid. That was the case in *McMaster University v. Wilchar Construction Ltd.*, a pre-*Ron Engineering* decision. As indicated earlier in this book, the judgment in *Ron Engineering* states that if the bidder's error had been evident on review of the bid, the result would have been dramatically different. It bears repeating. Reallocation of the risk of an error in the bid is limited to an invisible or latent error. An owner who attempts to "snap up" an apparent error will not succeed.

## Substantial Compliance/Irregular Bids Revisited

Most of the industry would define an "irregular" bid as one that possesses all the essential information required by the bid documents while it is, in some way, incomplete or mistaken. Examples of irregularities include unfilled blanks, omission of phone numbers and addresses, and obvious mathematical errors. Before *Ron Engineering*, and even after, a bid that was merely irregular was treated as compliant and admitted into bid evaluation except where the bid documents called for perfection.

As reported earlier in this book, the British Columbia Court of Appeal accorded "irregularity" a suitable post-*Ron Engineering* name — "substantial compliance." In the *British Columbia v. SCI Engineers* case, the bidder had submitted revised unit prices by fax just prior to bid closing. A submission in this media did not comply precisely with one of the conditions of the bid documents. Sorting out the validity of SCI's bid, the court relied upon Contract A as defined by *Ron Engineering* and concluded that perfection was not a requirement of these particular bid documents. In making the ruling for the Court, Chief Justice McEachern included the following comment:

> With respect, we think there are no circumstances in this case which requires the Crown to apply a strict rather than a substantial compliance test, particularly when the Crown was satisfied that no confusion was caused by the last revision.

Mr. Justice McEachern referred to a passage from *Ron Engineering* where Mr. Justice Estey addressed the same issue. The passage quoted read:

> It would be anomalous indeed if the march forward to a construction contract could be halted by a simple omission to insert in the appropriate blank in the contract the numbers of weeks already specified by the contractor in its tender.

So, it is fair to say that the threshold for compliance of any bid is, as long as the bid documents allow it, "substantial compliance." However, determining whether "substantial compliance" has been achieved or not is just as difficult as distinguishing between an irregularity, a qualification or an informality. Different label, same test.

Of all of the challenges to "substantial compliance," the biggest comes from issues dealing with price. Consider a bid where all the appropriate work divisions are listed, priced and correctly subtotalled. A number at the total line of the bid is mistakenly transposed to yield a bid price which differs from the sum of the division subtotals. Now what? What distinguishes an irregularity from something more serious is that the error and correction are *both* obvious. The example in this paragraph is an irregularity.

An example of a bid which would not be an irregularity is one where all of the divisions are listed, priced and subtotalled, but the bid price does not match the sum of the division subtotals and there is no transposition. The quick (and maybe wrong) answer to this is that the bid price must be the sum of the division subtotals. However, it could easily be a mistake within one of the divisions. The problem with this example (apart from being debatable) is that there is *more than one possible* correction. The prudent owner would rule this bid non-compliant.

A more concrete example, and one found earlier in this book, is *Vachon Construction*. The low bidder expressed one price in words, and a higher amount in numbers. The British Columbia Court of Appeal ruled that the bidder with contradictory pricing was non-compliant. The court reasoned that price was such an essential component of the bid that Contract A did not come into existence because the bid failed to clearly state the bid amount. Coming at it differently, there was more than one *possible* correction.

Considering the examples, one might think that the distinctions between them are without significance. Not so, given that Contract A either arises or not based on the state of the bid at the instant bidding closes. Taken in that light, the different examples and different results display some logic.

In *Ron Engineering*, Contract A arose because, at the instant of bid close, the bid was compliant in every respect. The owner learned of the error only later when the bidder raised it. Contract A bound the owner and bidder at the moment of bid close when the (apparently) compliant bid accepted the owner's offer to enter Contract A. The bidder was stuck when the owner then refused to release the bidder from its Contract A obligation not to withdraw its bid.

In the example given above where the division subtotals differ from the price bid due to transposition, it is apparent from the moment the bid is submitted what the bidder intended. There is no need to ask the bidder how to correct the error. Correction of the error works no mischief. The bid is compliant. In *Vachon*, it was not possible to know at the moment that bid was submitted which price had actually been bid. Something was clearly wrong and the contractor had to be asked to clarify.

Different bid documents could have changed the outcome in *Vachon*. For example, the bid documents could have provided that where there is a discrepancy between the price and words and the price in figures, the price in words must prevail. The Ontario Bid Depository Rules include just such a provision which reads:

> If words and figures are used and they do not agree, the words take precedence.

With the Bid Depository language in place, the need to ask the bidder a question disappears and the bid is compliant, although it may not reflect the intention of the bidder.

Most bid documents permit the owner to weigh irregularities — "*may*" is usually the operative word. Where the bid documents are silent, the custom of the trade is that an error must be minor in nature to constitute an irregularity, a flaw that might survive the test for "substantial compliance."

Wise is the bid-calling authority that carves itself out a discretion to deal liberally with irregularities, and thus create a pathway to "substantial compliance." After all, when was the last time an owner saw a set of perfectly completed bid documents?

## All Bids Are Non-Compliant

One might think that full-field non-compliance is the result of poor bid documents or the work product of a group of dysfunctional bidders. Oddly enough, all bidder non-compliance is often intentional.

At least one prominent industry association has mounted a campaign aimed at dissuading owners from issuing Supplementary Conditions which depart significantly from the CCDC/CCA family of standard form construction documents. The virtues/vices of Supplementary Conditions are the subject matter for an article, or maybe a book, but the result of the campaign has been full-field non-compliance on several large projects.

However it happens, the question remains: what should an owner do when all the bids it receives are non-compliant?

## *(a) What Does The Owner Have?*

When all bids are non-compliant, no bidder has Contract A with the owner because the bidders have not accepted the owner's offer to enter Contract A. So, the first part of the diagnosis is that the owner has no contractual obligations to the bidders governing bid evaluation and the Contract B award. Conversely, a bidder may avoid an award of contract if it alleges mistake before the owner accepts its bid (see the *Belle River* case in Chapter 1).

The bidders' end of the equation requires some analysis. First, have any of the bidders made an offer for Contract B (or something close to it) that is capable of acceptance? To determine this question, the owner must perform a set of tests which are similar to the measures it uses to determine compliance or non-compliance respecting Contract A. Here, the old measuring sticks — pre-*Ron Engineering* — for determining whether an offer is "formal" or not may be useful.

For example, has the bid been properly signed/executed by the bidder? Has the bidder made a clear and unequivocal offer to perform the scope of the work, or something close to it?

Assuming that an offer capable of acceptance has been made, is there anything missing from the offer that would persuade an owner to disregard it? For example, did the bidder submit the required bid security and agreement to bond? An owner could have a valid offer from a bidder (really a counter-offer). But, it may be far enough away from what the owner really wants that, although it is a valid offer, it is not one the owner would consider accepting.

## *(b) Evaluation: What Is An Owner To Do?*

So, what if an owner has an array of non-compliant bids that are valid offers, and attractive to the owner?

If an owner asked that question of the construction industry, the industry would say that the owner is "obliged" to deal with the lowest bidder. That might also be the owner's inclination. But, if the bid documents include the "privilege clause," neither the industry's wish nor the owner's inclination represent a requirement. Unless the bid documents commit the owner to an award to the lowest bidder, there is no legal obligation to make any award to the low bidder or to anyone else.

Some bid documents include a provision which permits the owner to negotiate with bidders after bids have closed. Bidders know (like it or not) that an owner has a right to negotiate and may exercise it if it so elects — Contract A or not.

To meet full-field non-compliance head on, some owners now include in their bid documents a more restrictive negotiation provision like the following:

> In the event that all bids are non-compliant, the Owner reserves the right to rebid the project or to conduct contract negotiations with a non-compliant bidder.

Most bid documents have nothing at all in them signaling that an owner will conduct post-closing negotiations with anyone, never mind non-compliant bidders. So, a revised question needs to be posed: when all the bids are non-compliant and the bid documents do not speak of post-closing negotiation, what should the owner do?

Assuming the owner has one or more offers for the project that are valid offers and do not contain any unacceptable features, the owner is free to negotiate with whichever bidder it might wish. The wisest and least controversial choice is to negotiate with the lowest bidder.

### *(c) Reject First and Then Negotiate?*

Most bid documents require that the offer made by the bidder be irrevocable for a period of time and backed by bid security. When all the bids are non-compliant, there is no Contract A as already noted. But, the owner still has an irrevocable offer backed with bid security. So, (theoretically) the owner still has something of value — an offer which cannot be revoked except for mistake. The owner should therefore conduct its negotiations without rejecting any of the bids.

Unless the bid documents specifically provide that the owner may negotiate with more than one non-compliant bidder at the same time, the owner should negotiate with one bidder at a time. If that negotiation proves fruitless, move on to the next most desirable bid, and start over.

### *(d) First Principles*

If there is no Contract A with any bidder, the owner has no *contractual* obligation with respect to the way that it runs the bid process. However, it has published bid documents which describe a process and which, in all probability, constitute representations by the owner of how it will behave in a commercial setting where bidders will spend money to make offers. A prudent owner will avoid conducting its bid process in a manner that conflicts with the express provisions of the bid documents.

Where the bid documents are silent on a point such as negotiation (remember: all the bidders are non-compliant), a prudent owner will determine what industry practice is, and follow it. If the owner departs in any material way from the process described in the bid documents, or from uncontradicted industry practice, it may find itself the defendant in an action for misrepresentation because one or more bidders may have relied to their detriment on the representations in the bid documents, and incurred the time and expense of preparing a bid.

Suppose for example that an owner states in its bid documents that it will consider offers only from bidders who carried pre-qualified mechanical and electrical trades. This owner might be sued for misrepresentation by a non-compliant bidder who lost the project contract having played (more or less) by the rules in its bid, only to learn that the owner awarded the contract to a bidder carrying mechanical or electrical trades which were not pre-qualified.

If there is any good news to an owner being sued, the measure of damages for misrepresentation is usually less punitive than contract damages (see Chapter 14). In this case, the best recovery the wronged bidder can make for a misrepresentation is to recover the cost of bid preparation, *i.e.*, to be restored to the position it would have occupied had the misrepresentation not been made. As suggested in Chapter 9, success in a misrepresentation case is no picnic.

Intentional non-compliance, as practised by some industry associations, is a blunt instrument. When all the bidders are non-compliant, the protection of Contract A is gone, and the owner finds itself in a pre-*Ron Engineering* landscape where it has very few legal obligations to the bidders. Some owners may play one contractor off against another.

A better approach is for owners and contractors to resolve their differences around Supplementary Conditions beforehand so that intentional non-compliance does not destroy either the fundamental purpose of bidding or the behavioral checks and balances which come with Contract A.

When all the bids are non-compliant, the dynamics strongly favour the owner — big time!

## EVALUATION ISSUES

Once the matter of compliance is out of the way, an owner moves on to the equally daunting task of evaluating surviving bids. Here, we have collected some of the challenging, and more frequent, bid evaluation issues.

## Pre-Qualification

Here, pre-qualification refers to the process carried out by an owner or its consultant to determine, among interested bidders, those who should be permitted to bid. The purpose of the exercise is to identify those bidders who have the skill, experience and resources to carry out the project. If pre-qualification is properly used, the owner is prepared to make an award to any one of them.

One of the virtues of pre-qualification is that it allows the owner to conduct an evaluation process outside the confines of Contract A. The submission of credentials by bidders does not create contractual obligations between the parties; therefore an owner is free to consider information about a prospective bidder during pre-qualification which might get it into trouble if the evaluation took place within Contract A.

For example, another owner might call and warn a colleague against a particular contractor. Because of the relationship between the owners, the owner receiving the information is prepared to rely on the advice without further inquiry. If the owner received the same information during evaluation within the confines of Contract A, the implied duty of fairness would require the owner to satisfy itself that the warning was justified.

The flip side of pre-qualification is the expectation of the bidders that the lowest compliant bidder will win absent some new and important commercial event that has taken place subsequent to the pre-qualification process. This simplifies matters for bid evaluation. At the same time, it restricts the field of movement for the owner.

As with any bid process, the bid documents describe the four corners of the evaluation process, with help from implied terms. Where pre-qualification has occurred, the bid documents may still be written so that the evaluation includes some kind of ranking between the bidders as to their skills and resources, schedules and related items — all amounting to a "nuanced" view of cost. This is not a popular approach among bidders who believe they have already run that gauntlet. And, for owners, it is a bit tricky since the elements that underlay the pre-qualification process are again gathered within Contract A and the owner may be called upon to account for, among other things, fair and equal treatment in the rankings it makes.

So, if pre-qualification has been employed, the lowest compliant bidder should win except under circumstances so extraordinary as to justify use of the privilege clause to reject the lowest compliant bidder. Or, the owner has provided discretion to take a "nuanced" view of cost — the concept which is described next.

## Nuanced View of Cost

A "nuanced" view of cost in a bid setting is not new. For many years, and long before *Ron Engineering*, owners have used more than bare price to determine the most attractive bid received. When owners have been challenged for analyzing more than just price, courts have upheld the owner's actions where they have related to schedule, quality of work and even claims history. (See *Elgin, Acme, M.J.B.* and *Sound Contracting*, previously reviewed in this book.)

A "nuanced" view of cost attained national recognition in *M.J.B.* where the Supreme Court of Canada found "particularly helpful" the description of a nuanced view provided by *Goldsmith On Canadian Building Contracts* which speaks of an owner discretion to

> consider not only the amount of the bid, but also the experience and capability of the contractor, and whether the bid is realistic in the circumstances of the case.

This discretion does not require express mention in the bid documents.

Within a year of the *M.J.B.* decision, the British Columbia Court of Appeal reversed the trial decision in *Sound Contracting* and, in doing so, extended the "nuanced" view of cost to include the claims history of the contractor even though claims history was not mentioned in the bid documents. What made this decision "worrisome" (a word used by Mr. Justice MacEachern who authored the decision) was that the City of Nanaimo had applied claims history to one bidder only, making the evaluation uneven.

In bid evaluation today, a "nuanced" view of cost is an appropriate measure when an owner approaches bid evaluation. The breadth of the nuance will always be governed by the terms of the bid documents. For example, bid documents that emphasize the importance of a completion date will bring a consideration of schedule, as a matter of cost, easily within the boundary of bid evaluation.

Similarly, a bid process which has *not* had the benefit of pre-qualification will include in the evaluation mix such elements as experience, skill, financial resources, claims history and workmanship — all leading to an owner conclusion of what represents the best-value compliant bid (rather than the lowest compliant bid) for an award.

Use of the "nuanced" view of cost will always be tricky because at least some elements of it will be debatable among the bidders.

The best approach for an owner is to list as "included" those components of a bid or a bidder's history that are important to it and will be considered during bid evaluation.

## Unrealistically Low Bids

Owners frequently receive bids on large projects at prices known to be less than the contractor's cost. With reason, owners are concerned that the contractor may be unable to complete the project, leading to a forced marriage between the owner and the bonding company. An equally troubling scenario has the low bidder opening its claim file on the day of contract award.

The owner receiving an unusually low bid will usually direct its consultant to study that bid carefully. There may be an apparent mistake, or there may be no mistake at all. If the owner has included a privilege clause, or carved out for itself the latitude evidenced in *Megatech*, it can reject a low bid and select another. Such a selection will provoke an outcry, and perhaps litigation.

Before rejecting any low (compliant) bid, the owner needs to carefully document its reasons in order to avoid the outcome in cases such as *Tarmac Construction* (formerly *Wimpey*), where the bald exercise of the privilege clause was punished.

Bids in two of the cases reviewed in this book could be called unrealistically low; however, the owner had chosen not to reject. The owner lost one case and won the other. In the *Ron Engineering* and *Ottawa v. Canvar* cases, the bids were judged compliant on their face, but immediately after bid opening, the respective contractors alerted the owner to significant mistakes which had led both bids to be well below that which the bidder intended.

Consider the owner's position if both of these contractors had signed instead of balked. In both *Ron Engineering* and *Canvar*, the bids were an order of magnitude below the second bidder. In *Ron Engineering*, the bid was close to the consultant's estimate, while in *Canvar* the bid was some 25% below it. In both cases, the evidence appears to have been that the pricing did not set off alarm bells. In *Canvar,* that turned out to be a stretch.

As a practical matter, an owner concerned about an unrealistically low bid might look at two obvious badges when making an evaluation.

The first badge is a bid which is below the margin of error stipulated by the cost consultant or design professional in its pre-bid estimate. Frequently, that margin is plus or minus 5%. A bid that is (say) 10% to 15% below the pre-bid estimate might raise an eyebrow. Why just one eyebrow? Because at this point, the owner does not know whether the consultant is wrong or the contractor is wrong.

The second badge is what the other bidders have done. Ideally, an owner likes to see tight pricing grouped around the pre-bid estimate. If the low bidder is also significantly below the other bidders (assuming they are competent and interested), the other eyebrow goes up too.

If both eyebrows are up (so to speak), then the owner may be confronted by different scenarios. The first is a bid that is unrealistically low and would justify rejection under the privilege clause. The second may be an error on the face of the bid which would notify the owner of a mistake that a court would not let the owner "snap up."

Some experienced contractors have suggested that this "two eyebrow" approached is too simplistic. They tell stories of having bid way below other bidders and way below the pre-bid estimate, and of having still done "all right" on the project. How can the owner tell? To most rules, there are exceptions. The question is whether the owner wants to bet the long shot (the exception) or opt for the more prudent choice (the rule).

If the case law on invisible/latent bid mistake reflects the current ethic of the marketplace, then the owner will "sock it" to the mistaken contractor. That hardly seems like the fair thing to do if the owner's expectation was to pay a price in the neighbourhood of the consultant's pre-budget estimate.

As a post-type script, the *Calgary* case tells us that it is not "okay" to allow the contractor to adjust its bid to a higher number which will still be the low bid. No doubt about it, such a post-bid amendment would breach Contract A with compliant bidders.

## Unbalanced Bids

The "unbalanced" bid is one where the contractor artificially inflates the value of divisions of work to be done first. The industry inelegantly describes this as front-end loading.

The lack of balance may be evident in a bid, or in a required after-bid submission. The result for the contractor is a cash flow which runs ahead of project completion. For the owner, this practice raises (at least) serious construction bonding problems (the bonding company may be partially or entirely discharged) unless it can be prevented. It also raises problems with lenders who are not anxious to have their loan advances running ahead of the value of construction.

Bid documents often reserve to the owner the right to reject, limit or modify the contractor's allocation of the contract price to the divisions of work in its bid. While these provisions do not allow the owner to modify the bidder's price, it may permit the owner to redistribute the line items underlying the price.

The Ontario Ministry of the Environment is a bid-calling authority which, over the years, developed extensive Instructions to Bidders that enabled it to deal with a variety of bid issues, including unbalanced bids. In a version of its standard form bid documents, which was in use in the late 1980s, the Ministry gave itself the discretion to reject an unbalanced bid. The bid documents went on to limit the amount a contractor was permitted to carry for mobilization to 10% of its bid price.

However, the most interesting provision was a clause which enabled the ministry to change the values on the contractor's line items to correct what it saw as unbalanced prices. The contractor was obliged to accept the Ministry's view of balance.

Where a bid-calling authority does not anticipate an unbalanced bid in its bid documents (which include the construction contract), it may find itself pushed to disqualify a bid it would prefer to accept. Disqualification for lack of balance would probably be a good business reason if it is exercised in conjunction with the garden variety privilege clause which reads, more or less, "The lowest or any tender not necessarily accepted." Most owners would prefer to have explicit options, and the bid documents can create them.

## After-Bid Amendments By Contractors

In the *Acme* and *Calgary* cases, the courts held that an owner had no duty to permit a mistaken bidder to increase its bid price to correct a hidden mistake even if the adjusted price would still be the lowest price.

The flip side of this argument is that it would be a violation of Contract A for the owner to permit an after-bid amendment when the bid documents do not expressly permit one. An after-bid adjustment would favour one bidder over the others. If the lowest bidder can change its bid after all prices are known, why should not the higher bidders have the same opportunity? The courts signalled in the *Acme, Calgary* and *Vachon* cases that owners should not allow this kind of after-bid activity.

## After-Bid Amendments By Owners

On the opening of bids, owners sometimes find (unhappily) that all the bids are well over budget. There seems to be general agreement throughout the industry that the pathway to a satisfactory bid is set out in both CCDC-23 and CCA-29. The more extensive statement on the point comes from CCDC-23 (1982) and reads as follows:

> In the event that all bids received exceed the owner's budget, the owner should negotiate changes in the scope of the work with the bidder submitting the lowest acceptable Bid. When the negotiations result in a Contract Price acceptable to both parties no re-bidding of the project is necessary and the Contract should be awarded at the negotiated price.

If negotiations fail to produce a Contract Price acceptable to both parties, or if, in the first instance, the changes contemplated result in a value in excess of 15% and the owner wishes competitive prices thereon, the Bid Documents should be amended and invitations to re-bid should be restricted to the bidders who submitted the three lowest acceptable Bids on the original Bid call. The Subcontractors invited to re-bid should likewise be restricted to those who submitted the three lowest acceptable prices in each Contractor's original Bid.

Controversy follows the owner who approaches its budget problem by convening an auction. A "post-tender addendum" signals the commencement of the auction when issued to more than the lowest bidder with minimal changes in work scope.

An equally unpalatable approach to a budget problem involves the owner that seeks recommendations from the lowest bidder on how to reduce the price to meet budget. Once it receives the information, the owner "shops" the recommendations to one or more of the higher bidders. This approach to the lowest bidder starts out like the process outlined in CCDC-23 and then mutates into bid shopping. If there is nothing in the bid documents permitting bid shopping, this shopper is probably in breach of Contract A.

If the bid documents do not state what course will be followed if the budget is exceeded, Contract A will be breached if the owner adopts a procedure that is significantly different from the one suggested by CCDC-23. Unless the bid documents permit all bidder negotiations, it should not happen.

In *Midwest*, the owner created bid documents which permitted post-closing negotiations with bidders who were non-compliant. When the negotiations with Midwest proved fruitless, Midwest claimed that the act of negotiation brought Contract A into being. As an alternative, Midwest argued that, even without Contract A, the owner owed it a free-standing duty of fairness which it had breached. The British Columbia Court of Appeal ruled that Midwest had no case on either count. The mere act of negotiation did not bring into being Contract A when the owner's offer of that contract had been rejected, and there is no freestanding duty of fairness.

More recently, a British Columbia court went the other way in *Kinetic*. The bid documents permitted the owner to "retain" a non-compliant bid, which the owner did. Having evaluated the non-compliant bid, the owner selected another bidder and faced litigation. The motion's court judge ruled in favour of Kinetic, reasoning that Contract A came into being when the owner retained Kinetic's non-compliant bid and evaluated it. What flowed from Contract A was that the owner owed Kinetic a duty of fairness which it breached by awarding to another bidder.

The *Kinetic* decision is under appeal as this book goes to press. It will be interesting to see if the *Kinetic* decision will stand, given the findings of the British Columbia Court of Appeal in *Midwest Management* and those of the Supreme Court of Canada in *Martel*, both of which have held that there is no freestanding duty of fairness outside Contract A.

In *Graham*, the British Columbia Court of Appeal looked at *Kinetic* but disregarded it. The decision in *Graham* was that a "discretion clause" included in the bid documents did not "operate" unless Contract A came into being. The conflict between *Kinetic* (fairness may exist without Contract A) and *Graham* (clause inoperative unless there is Contract A) will cause confusion in the industry until another court steps in to resolve it. Meanwhile, if you are a bidder, don't expect to be entitled to fairness unless you have Contract A. And, expect the clauses in bid documents — which form part of the offer for Contract A — to operate, even during assessment for compliance.

Another approach which at least one multi-national owner has used is called "target pricing." In this technique, the owner receives all bids, opens them and then communicates to one or more of the bidders that it had a lower price in mind than the one submitted. The implication is that if the bidder is prepared to meet that "target price," it will be awarded the job. This approach has been used even when the bids received are within the pre-bid estimate. Whether it is the intent or not, the result of "target pricing" is an auction which is not acceptable to the construction industry. The practice is a breach of Contract A unless it is provided for in the bid documents.

## Unsolicited Alternatives

The contracting fraternity is entrepreneurial and imaginative. Often, they see opportunities to improve on an owner's design package, or a bidder tries to get a "leg up."

Where a bidder provides an unsolicited alternative, and changes the bid form or other mandatory submission to do so, it usually creates non-compliance. That is so if the bid documents prohibit alteration or deletions. Not a smart move when bidders make it.

Some bidders provide unsolicited alternatives by including something extra in the bid package, such as a letter describing the alternative and the implications for price and/or schedule. If the bid documents prohibit such a submission, the bid is non-compliant.

Where the bid documents are silent on unsolicited alternatives, and the bidder, having submitted a compliant bid, provides an alternative by way of an additional submission, the owner has a number of options.

First, the owner can disregard the "extra" and deal with the bid it has received.

Another option, and a tricky one, raises its head where the unsolicited alternative appeals to the owner. If the owner includes the unsolicited price/schedule information in the overall bid analysis, it probably breaches its duty of fairness to all the bidders who have Contract A with the owner. This is not much different than *Chinook*.

If the owner gets lucky, the bidder with the intriguing unsolicited alternative is also the lowest compliant bidder. If that is the case, the owner may carefully proceed to award the contract to the lowest bidder and then issue change order no. 1 capturing the advantage of the unsolicited alternative.

Some bid documents actually provide for unsolicited alternatives and describe how they are to be submitted. When bid documents go in this direction, the bid documents often describe how the owner will evaluate unsolicited alternatives. This is a dangerous area for several reasons. First, the value of the unsolicited alternative has not been tested against the market. Second, if unsolicited alternatives are included within the terms of Contract A, the owner is accountable for how they are handled. Third, encouraging unsolicited alternatives may make life more complicated for the consultant and more expensive for the owner.

Tricky business, unsolicited alternatives. If the project has that much room for innovation, the better approach is to engage a contractor during the preparation of working drawings and flush out alternatives which can be priced to the full field of bidders.

## Owner Errors/Omissions in Bid Documents

The Supreme Court of Canada decisions in *Edgeworth* and *Auto Concrete Curb* send a clear message that, under the right circumstances, a contractor who suffers a loss due to errors/omissions in the bid document may pursue the owner's consultant directly.

The *Edgeworth* case extended the scope of the risk being assumed by consultants who routinely prepare drawings and specifications. Before *Edgeworth*, the perimeter of the consultant's concern was limited to the situation where, having prepared the drawings and specifications, the consultant assisted with the calling of bids and administered the resulting contract. No more.

In *Edgeworth*, the sole duty of the consultant was to prepare the drawings and specifications. It was years later that the owner used them to build a highway. The engineer had no role (and received no fee) in the calling of bids or in the execution of the work. The Supreme Court of Canada found that the consultant's duty extended to a contractor it knew nothing about, on a project it no longer influenced. Design professionals are troubled by the decision; contractors wonder why it took so long.

This court-imposed duty has led to new provisions in the contracts between owners and their consultants, limiting the liability of the consultant on the drawings and specifications. At present, the courts are using tort (negligence) theory to create a connection between consultants and contractors, concluding that any consultant should have the ultimate user of its drawings and specifications in mind when the consultant issues them.

When the Supreme Court of Canada overturned the Court of Appeal for Ontario in the *Auto Concrete* case, all design professionals breathed a sigh of relief. The Court of Appeal had upheld a trial decision which found the consulting engineer negligent for failing to anticipate a construction method the contractor proposed to use, and for failing to provide the contractor information in the bid documents about the kind of permits that this method would require.

These lower-court findings were made in spite of express language in the bid documents making the contractor responsible for selecting its method of construction and for obtaining the requisite permits. The Supreme Court of Canada held that it was not the consultant's job to anticipate in the bid documents all the ingenious ways that a contractor might elect to execute its work. In this case at least, neither owner nor consultant was liable.

Let us now compare *Auto Concrete* and *BG Checo*, which are studies in contrast. The former case illustrates a situation where the assignment of risk clause (that is, when the contractor selects the method) effectively shielded the owner from liability in contract and the consultant from liability in negligence. The *BG Checo* case held that the global assignment of risk clause was inapplicable where the owner tried to apply it to negate a specific contract term.

In *BG Checo*, the bid documents told interested contractors that a Hydro right-of-way would be cleared by others. When the contractor conducted its pre-bid inspection, it observed that the right-of-way was a mess. Relying on the bid documents, the contractor ignored the mess. The owner did not clear the right-of-way. This shifted the expense to the contractor, and caused a schedule delay.

When the contractor made a claim, the owner pointed to a clause which assigned the contractor the risk of any site condition that the contractor observed during its pre-bid inspection. The decision in favour of the contractor means that a specific instruction by the owner to the contractor (owner will clear the right-of-way) cannot be negated by a blanket exclusion (contractor responsible for site conditions). If there was any ambiguity between the two (and most would suggest that the contractor's interpretation was clearly correct), that ambiguity was construed against the owner/draftsperson — B.C. Hydro.

The *BG Checo* case was about a conflict in the bid documents. *Mawson Gage* was a case of owner mistake where the owner compounded its error in the bid documents by misinterpreting *Ron Engineering* and exercising leverage that it really didn't have. The owner scared Mawson Gage into taking a subcontract in the first skirmish, but eventually the owner lost the war.

In *Mawson Gage*, the injured party was a subtrade. The injury resulted when the owner assembled a set of bid documents with an entire section missing. Mawson Gage received that set. Thus, its subtrade bid to the prime contractor did not include the missing work. The omission was discovered after prime bids closed.

When the owner was told, its response was to threaten to call the contractor's bid security if the contractor did not force its sub to perform. In response to the owner's hard line, everybody signed, and the sub performed the missing work without any guarantee of payment. As it turned out, the owner's action only put off the inevitable consequences of its own negligence. Its attempt to "get something for nothing" was unsuccessful, and the sub was paid for the work it had done.

## Subcontractor Error in Bid

Errors by subcontractors are treated in the same way as errors by general contractors. Contract A between a contractor and a subcontractor comes into being when the contractor elects to "carry" (that is, use) a trade's bid in putting forward its own bid to the owner.

In different cases, "carry" has ranged from listing the name of the trade in the prime bid (very clear) to incorporating the trade's line item price without including the name either in the prime bid or in the contractor's own working papers (much less clear). But when Contract A is found to have been created (which would be in the vast majority of cases), the terms are derived in the manner described in *Naylor*. This is to say that the terms of Contract A are defined by the particular set of prime bid documents upon which the subcontractor bids are based.

Subtrade errors arise most frequently when the bid depository is not used. Trade prices are taken at the last minute, usually by telephone. In the heat of the moment, it is easy for mistakes to occur on either side. That is what happened in *Scott Steel*.

The court in *Scott Steel* had to determine whether the contractor was obliged to do the work using the trade which had made an error. Before the error had been discovered, the contractor had listed the trade in its prime bid. That trade was still in the prime bid when the owner accepted the bid. The court looked for a way to relieve against a commercially unacceptable result.

It was in this setting that the Ontario Divisional Court articulated the finding that the subcontract (Contract B) had not come into existence because the contractor had not expressly communicated its acceptance of the subtrade bid after being awarded the prime contract. This was the right result — for the wrong reason. Troubling was the feeling that establishing such a rule would encourage contractors to position express communication of acceptance against price concessions from trades.

The principle enunciated in *Scott Steel* has been modified in the Supreme Court of Canada decision in *Naylor*. The Supreme Court of Canada confirmed that it is an *obligation* of the contractor under Contract A to communicate acceptance of the subcontract to the carried trades within a reasonable time of receiving an award of the prime contract. Also, bid shopping, thanks to *Naylor*, is no longer merely anti-social. It is a breach of Contract A.

But what happens when a sub refuses to honour its mistaken bid? When the subtrade's confirmation of its oral quote backs away from that quote, the contractor which carried that sub and won the prime contract is in a jam. If the sub refuses to honour its oral quote, it will usually be liable to the contractor if it walks away. That usually means a trip to court.

While the legal principles which apply to subtrade bid cases are fairly clear, the evidentiary situation is often not. Last minute quotes are frequently oral, are jotted in a journal or are transcribed right into a bid. It can be anybody's guess as to who actually said what to whom.

Courts apply legal principles to fact situations. Because the close of subtrade bids (outside a bid depository situation) is so fluid, the trial judge has to cut through conflicting evidence to make a determination upon which to apply the law. The facts that can be proven in a courtroom frequently do not match what has actually happened. The trial judge only learns what the rules of evidence will allow him/her to learn.

Where the bid depository is used and a subcontractor makes an error, the same principles apply, but with a twist. The twist is the bid depository rules themselves.

Typically, the bid depository calls for the trade bids to close two working days before the prime bids close. This prevents the last minute mayhem of telephone quotes. If a subtrade discovers an invisible error *after* the prime bid has gone in, it has to decide whether to perform the work under its mistaken bid (taking on an almost certain loss) or refuse to do the work and get sued (taking on an almost certain loss plus the joy of paying its own legal expenses and those of the contractor).

If the mistake is discovered *after* the bid depository for a particular project has closed but *before* the subtrade bids are released to the prime bidders, some bid depositories will permit the sub to withdraw. In the Ontario Bid Depository Rules, subtrade prices are released to the prime bidders two hours before the prime bid is due to close. If a withdrawal happens before the prime bid deadline, all the contractors that were planning on using that subtrade bid will be in the same boat and will move to another subtrade bidder that is willing to have its price carried by that prime bidder. No harm, no foul.

## Tied Bids

It does not happen very often but it does happen: identical prices from two compliant bidders. Bid documents rarely provide for a tie. CCDC-23 does not provide any help; CCA-29 does.

CCA-29 deals with "tied bids" as follows:

> In the case of tied bids, the contract should be awarded on the basis of the most advantageous time schedule. If a decision still cannot be reached, the Owner should determine the successful bidder by coin toss in the presence of the tied bidders.

What if one of the three parties says "no" to the coin toss?

The owner can probably navigate these difficult waters by performing the coin toss and awarding the contract accordingly. If the bid documents have not provided for the situation, a court dealing with such a case would have to imply a term. It would probably look to industry practice for that term and probably wind up singing from the CCA-29 song sheet.

As an alternative, the owner could go back to its bid evaluation. If the time schedule does not make one bid more desirable than the other, the owner may find, for example, that one of the bidders is carrying a subtrade or two with which the owner has had an unhappy experience. Stated generally, what the owner is looking for is an objective, business-related reason drawn from the bid evaluation criteria to prefer one bid over the other. If this revisitation is unsuccessful, go to Lady Luck.

At the time of writing, there does not appear to be any case law dealing with identical prices between compliant bids.

## Electronic Bidding

The submission of bids electronically is not in the mainstream — yet. There are some owners, such as the Government of Canada, who post their invitations to bid and the bid documents on a website. MERX is the most prominent Government of Canada example. Even with MERX, the bid documents, complete with signatures and bonds (where required) are delivered in "hard copy."

Although an enormous amount of world commerce is carried out over the internet, bidding in Canada presents some interesting challenges. There is reluctance on the part of consultants to see their valuable and copyright protected work product appear on a website to be downloaded by all comers and, perhaps, not used for the purpose intended.

There is also the impediment of the medium and the surrounding legislative framework.

If a bid is to be submitted on or before a certain time, what happens if the owner's website is down? What happens if the bidder sends before the bid period ends, but the owner receives it after? How are bidders protected from having third parties gain access to their confidential bid information? And the list goes on.

Several provinces have put in place legislation to govern electronic commerce. Because the legislation is intended to protect a wide range of buyers and sellers, some sensible consumer provisions make bidding problematic. For example, if the legislation gives the accepting party a period of time within which to reconsider or correct the offer, how does that impact irrevocability and the formation and performance of Contract A?

There are examples of owners using the internet to augment a bid process. One such example calls for bidders to submit a bid (hard copy) by a specific closing time. Then, the bid documents provide that, say, the next morning at 9:00 a.m., all the prices bid will be posted on the website. Each bidder has the opportunity to amend its own price one time. The low bidder is determined at the end of this electronic auction. Structured correctly this process could create a Contract A/Contract B model envisaged by *Ron Engineering*.

Will electronic bidding ever become the norm? Probably. Based on the current state of the law in Canada, whether a bid is submitted by the internet or by carrier pigeon, the same law will apply.

## WHERE TO FROM HERE?

It would be comforting to assure our readers that we have captured all of the troublesome issues related to compliance and to bid evaluation, or to a mixture of the two. Bidding being what it is, the owner and contracting fraternity will come up with situations that neither the courts nor ourselves have encountered. Using the principles now well established around bidding law in Canada, the owner and the bidders, assisted by their lawyer, have the tools to puzzle their way through what bidding dynamics have created.

The puzzling through will always be made easier by bid documents which treat Contract A like a contract and include, as best we are able to include, terms which make the process competitive, transparent and fair. And, by so doing, achieve one of the policy objectives of the Supreme Court of Canada as expressed in both *Ron Engineering* and *Martel*, to promote and protect the integrity of the tender system. More about this in Chapter 14.

# Chapter 13

# DAMAGES AND OTHER RELIEF

## GENERAL PRINCIPLES

So far, we have dealt with the question of liability in bidding cases. However, proving liability represents only half the battle. We now turn our attention to the question of *damages*.

Damages are a sum of money awarded by a court as compensation for a breach of contract or a tort. Generally speaking, a party that suffers losses at the hands of another is entitled to compensation. The theories which underlie damages in tort and in contract will be discussed later in this Chapter. Right at the start, we wish to address the necessary connection between a wrong and the recovery of damages — proximity or its inverse, remoteness.

It is not just or practical to award damages for *every* consequence, however unusual, which may flow from a breach of contract or from a tort.

*Cornwall Gravel Co. Ltd. v. Purolator Courier Ltd.* is a case, not previously summarized in this book, which deals with the issue of proximity/remoteness and foreseeability.

In a nutshell, Cornwall prepared a bid for a construction project in Toronto. The bid submission deadline was 3 p.m. on October 2. The day before, Cornwall contacted Purolator to advise that it had a pick-up for Toronto that evening.

The bill of lading covering the shipment included this disclaimer:

> Maximum liability $1.50 per pound unless declared valuation states otherwise.

Cornwall stipulated that the bid should be delivered at 12 noon on October 2 rather than 3 p.m. — just to be safe. The driver who picked up the package was told that it contained a bid which had to be delivered on time.

On the way to Toronto, the Purolator car broke down. The bid was delivered at 3:17 p.m. and was not accepted. Had the bid arrived on time, Cornwall would have been the lowest bidder and would have been awarded the contract. The job would have earned Cornwall a profit of $70,000.

Cornwall sued Purolator which defended itself by pointing at its limit of liability of $1.50 per pound. Cornwall argued that the limitation of liability only applied to the value of the package and not to the consequential loss. As well, Cornwall argued that, since it told the driver what was in the package, the loss was foreseeable.

The court referred to a decision in the House of Lords in *Hadley v. Baxendale,* decided more than 150 years ago. Hadley owned a mill and hired Baxendale to deliver a broken mill shaft to a manufacturer who would fashion a new shaft. Baxendale promised to deliver the shaft the next day but through his negligence, the shaft was delayed in transit. Hadley claimed for loss of profit caused by the fact that the mill was out of service for a longer period than it should have been. This happened because of the delay in delivery of the shaft.

In deciding between Hadley and Baxendale, the court looked at two types of damages. One was general damages which are awarded for losses that the law will presume are the natural and probable consequence of a breach. If Baxendale had lost the mill shaft he would have been liable for the value of the shaft as *general damages.*

Hadley's loss of profit was another matter. These are *special damages* which are awarded for losses that are not a direct consequence of the breach so require special proof. Such damages are recoverable if the party which caused the loss could have reasonably foreseen, at the time of contract, that the loss was probable if the party failed to carry out its contractual obligations.

In the end, the House of Lords found that Baxendale did not know the special circumstances in which Hadley found himself. So, the loss of profit was not foreseeable to *both* parties at the time that the contract was made. The court refused to award Hadley the special (consequential) damages. (The lesson: If you are about to enter into a contract, and there is a possibility that a breach may lead to consequences and losses that are not immediately foreseeable, make sure that the other party is made aware of them!)

In the context of the *Cornwall Gravel* case, the question became whether Cornwall's loss was foreseeable to Purolator when the parties made the contract. Purolator, through its driver, knew that the package contained a bid and that it had to be delivered by 12 noon on October 2. The driver must have realized that the bid would be worthless if delivered late. And late it was, by some 3 hours and 17 minutes.

The trial judge determined that Cornwall had, in fact, communicated the special circumstances of the contract of Purolator as well as the damage which would flow from the breach of contract.

Before the court could make an award to Cornwall, it had to determine whether the limitation of liability — $1.50 per pound — applied. The judge found that the clause did have application but only to loss or damage to the package itself. The clause did not cover a consequential loss due to delay and Cornwall was able to recover its full damages of $70,000. The outcome at trial was upheld in the Court of Appeal for Ontario, and in the Supreme Court of Canada.

Two things flow from *Cornwall Gravel* which will be of assistance in the rest of this Chapter.

First, damages suffered by a party must be reasonably *foreseeable* by the other party in order to be recoverable.

Second, a clause which limits damages recoverable on a breach of contract will not protect the party attempting to rely on it unless that clause clearly identifies and specifically excludes the precise damage that was actually suffered.

## PROOF OF DAMAGES

A party sometimes expends all its effort proving liability — only to wake up to the sad conclusion that it cannot prove damages. So, having won, the party actually loses. Let us now examine the issue of *proof of damages*.

*BG Checo* was a Supreme Court of Canada case where the fight between owner and contractor included concurrent claims for damages for breach of contract and for the tort of misrepresentation. The Supreme Court of Canada confirmed that the two claims could be brought simultaneously. And, in the course of its reasons for judgment, the court provided a simple definition of how damages are measured for breach of contract and for the tort of misrepresentation:

*Contract:* The plaintiff is to be put in the position it would have been in had the contract been performed as agreed.

*Tort:* The plaintiff is to be put in the position it would have been in had the misrepresentation not been made.

But measurement of the damages is not the entire story. As we saw in the *Cornwall Gravel* case, there needs to be a relationship between the damage suffered and the event giving rise to those damages. That relationship has been described as *proximity* or *foreseeability*.

In contract, the test is whether the damage was in the "reasonable contemplation" of the parties at the time the contract was signed. As you can imagine, some things are, and some things are not.

One way of illustrating the principle is in a contract setting: The contractor is building an office building for the owner, and the owner requires it for a specific date. The contractor is six months late, and the delay is entirely the contractor's fault. The late delivery of the building means that rentals start six months late. These are damages that were in the reasonable contemplation of the parties, and the contractor will pay them unless the contract says otherwise.

There could be other damages as well. In *BG Checo*, the owner's misrepresentation meant that the contractor had to absorb the cost of clearing the hydro right-of-way and the losses flowing from extended contract duration. Those two kinds of loss were foreseeable, and the owner paid them.

Although a wronged party can pursue both a contract and a tort remedy in the same action, it cannot recover damages for both. The reason for the two-pronged attack is that one of the prongs may not succeed.

In the Contract A/Contract B world created by *Ron Engineering*, concurrent claims have often been made (examples: *Martel* and *Midwest*). The courts have been willing to use either theory of liability to discourage commercially unacceptable behaviour.

In this section, we will review some of the leading cases where owners paid damages to contractors (and vice versa) and where contractors paid damages to subcontractors (and vice versa). We will also look at some remedies where the relief is something other than damages.

Damages paid to contractors by owners for breaches of contract have been the most elusive. The first question to be asked is which contract has the owner breached? In all of the cases referenced, the answer is "Contract A". That can take the contractor anywhere from zero damages recovered to the profit it was likely to have enjoyed had it been awarded Contract B (often called *expectation losses*).

To illustrate the differences in possible recovery, assume an owner has received eight bids. The low bidder is non-compliant so that Contract A does not come into being. The bid documents contain no language which would allow the owner to waive the non-compliance. Owners being owners, the work goes to the non-compliant low bidder. Contract A has been breached — with the compliant bidders. Both the second low bidder and, say, the sixth low bidder sue the owner and have their trials heard one after the other (to get consistent findings of fact).

In both cases, the court agrees that the owner has breached Contract A. It applies the test in *M.J.B.* to find that the bid documents, properly construed, include an implied term that the owner will not award Contract B to a non-compliant bidder. Both Bidder No. 2 and Bidder No. 6 sue for a breach of Contract A, seeking loss of profit and, in the alternative, the costs of preparing their respective bids.

On the question of damages, Bidder No. 6 goes first. It argues that but for the breach of Contract A by the owner, it would have been awarded Contract B and should have its reasonable profit as damages. It loses this argument. After all, it finished sixth. Then it argues that it should recover the cost of bid preparation. The court is sympathetic; after all, the owner did breach Contract A by awarding Contract B to a non-compliant bidder.

Unfortunately for Bidder No. 6, this results in an award of one dollar. The court holds that if Contract A had not been breached, the lowest compliant bid would have received the work and Bidder No. 6 would have been in the same position, breach of contract or no. At this point, Bidder No. 6 kicks itself for not pleading misrepresentation (tort). The court might have accepted the proposition that the bid documents include a representation by the owner not to award Contract B except as described in the bid documents. That duty is owed to all bidders, and it was breached when the owner awarded to a non-compliant bidder — a misrepresentation. Bidder No. 6 might have been able to persuade the court that its damage was the cost of bid preparation — a cost incurred in reliance on the representations in the bid documents.

Bidder No. 2 is concerned with the turn of events that befell the argument of Bidder No. 6. It refers to the trial evidence which suggests that in the normal course of events, had the job not been awarded to the non-compliant bidder, the owner's purchasing policies would have required it to award the job to the low compliant bidder. Bidder No. 2 argues that on these facts, the balance of probabilities says it would have had Contract B. The court agrees and rules that Bidder No. 2 should recover its reasonable profit. The claim for the cost of bid preparation falls by the wayside. Bidder No. 2 enjoys a significant pay day without the trouble or risk of having to carry out the work. Same breach, very different results.

In the cases described below, the courts wrestled with their discomfort with Contract A. Judges recognized the right of a party to bring concurrent actions for breach of contract and tort. Some trial decisions have elaborated novel non-contract duties between owner and subcontractor (see *Twin City Mechanical v. Bradsil* and *Ken Toby*) and between contractor and subcontractor (see *Naylor*). Appellate courts have recognized that there is a duty of care but have not faced a case where the court felt that duty was breached — so far.

## BREACH OF CONTRACT A

### Awards In Favour of Owners Against Contractors

The first case ever where Contract A was breached was the great-grandfather of this jurisprudence, *Ron Engineering*. The breach was that the contractor refused to sign Contract B. The Supreme Court of Canada ruled that the owner was entitled to cash the contractor's bid bond. This did not cover the entire spread between the *Ron Engineering* bid and the second lowest bid which the owner accepted. As a result, the owner was not compensated for part of its loss. Nonetheless, it was the result which the bid documents provided.

The cases subsequent to *Ron Engineering* have proceeded differently. When a contractor breaches Contract A, that breach has a direct and immediate effect on the contract price which the owner must pay for Contract B. The owner recovers the spread between the defaulting low bidder and the second lowest bid. Whether it is fair or not, the measure is simple.

This simple measure of damages was applied in the *Calgary* and *Acme* cases. It was also applied in the trial decision in the *Ottawa (City of) Non-Profit Housing Corp. v. Canvar Construction (1991) Inc.* case, but then reversed on appeal. In none of those cases were the damages limited to the bid bond; the bond only served to provide some liquidity and/or security to the owner.

Once the breach of Contract A is found, it is relatively easy to determine what amount of money will put the owner in the position it would have occupied without the contractor's breach.

## Awards In Favour of Contractors Against Owners

When a court finds that an owner is in breach of Contract A with a contractor, the court must look further before it can decide how to measure damages. But for the breach, was it a given that the contractor would have been awarded Contract B? Sometimes yes, and sometimes no.

Where the breach of Contract A is judged to have resulted in the loss of Contract B, the contractor is usually awarded the profit it would have earned on the lost project. Because bidding and contracting is more art than science, the court takes the contractor's anticipated profit figure with a grain of salt. Typically, three discounts may be considered by the court:

- The first is to give effect to the possibility/probability that Contract B would not have been awarded to the contractor (maybe the chances were good but not absolutely certain).
- The second is to allow for other circumstances, such as unanticipated negative project conditions.
- The third addresses whether the contractor took steps to reduce (mitigate) its losses by seeking replacement work.

In the *Kencor* case, the Province of Saskatchewan was held to have breached Contract A. It was clear, the court reasoned, that without the high-handed behaviour of the Province (which overruled a staff recommendation), Kencor would have been awarded the work and earned a reasonable profit. The damages were agreed at $180,000, which "represents among other things, the plaintiff's loss of profit for not having been awarded the contract".

The *Colautti Brothers* case was one where the owner breached a duty of fairness implied into Contract A. However, the court did not conclude that the breach led inevitably to Colautti losing Contract B. The damage award was limited to Colautti's cost of bid preparation.

It is worth examining the extenuating circumstances which led to the *Colautti* award. The owner was the City of Windsor. It had a policy of awarding contracts to the lowest qualified (meaning capable) bidder. And Colautti was qualified. The problem was that Colautti's was the only bid received, and the City had a dilemma as to whether to open it or not. It elected to open the bid, thus making the Colautti's price public.

The City then found that Colautti was 40% over its budget and, when negotiations with Colautti failed to produce a satisfactory contract price, it neither accepted nor rejected the Colautti bid but re-tendered. Colautti responded to the second bid call by suing.

The court held that Contract A had come into being and that it had included an implied duty of fairness. The City breached that duty when it failed to follow the CCDC procedure for owners receiving but one bid (send it back unopened).

The court went on to hold that the privilege clause gave the City the right to reject the Colautti bid or any other, hence there was no linkage between the breach of Contract A and the recovery of the anticipated profit on Contract B. When the court awarded Colautti its costs of preparing the bid, the thinking was that Colautti was disadvantaged when its pricing was made public. Although the court found that the City acted in good faith and was entitled to reject the Colautti bid, it nonetheless awarded damages for the breach of the duty of fairness.

Perhaps, a more appealing analysis is that while the City might have been more prudent to return the bid unopened, the result would have been the same: Colautti would not have got the job. It was no one's fault that Colautti was the only bidder. All bidders take the risk that they might lose.

If Contract A does not arise until a bid is judged compliant and received for evaluation, how could the City have breached a Contract A duty of fairness by merely opening Colautti's bid? When the bid was opened, Contract A arose, and the court found that Colautti was treated fairly and in good faith. Makes a person wonder why the court awarded damages against the City?

*Thompson Bros.* is useful to this discussion because of the steps the court followed in awarding damages to Thompson. The court had implied a duty of fairness into Contract A, and then used a breach of that duty to prevent the owner from relying on the privilege clause to avoid liability. The court then had to determine what damages flowed as a logical consequence of this breach.

First, it asked whether Thompson would have received Contract B but for the breach. The court concluded that it was a "virtual certainty" based on the activities of staff before city council became involved.

Second, it looked at the profit which Thompson would have earned on Contract B to determine whether there were any job factors present which would support a discount. The court looked at several other cases where discounts were applied and decided none was warranted. Thompson was awarded its full estimated profit of $88,000. The trial judge did not turn his mind to the third potential discount — replacement work.

In *Thompson*, as well as in the cases cited in the reasons for judgment, a distinction was made between damages for breach of Contract B and damages for breach of Contract A. If the court is dealing with a breach of Contract B, it need not go through all the reasoning which the court in *Thompson Bros.* did. Contract B damages involve only a calculation of what profit would have been earned if the project had been carried out reasonably.

Where Contract A is breached, the award is the profit that the contractor would earn *if* Contract B is awarded. Because the probabilities of an award in *Thompson Bros.* were a "virtual certainty", the court did not apply any discount to cover the possibility of no award. In other fact situations, where the likelihood of a contract award is high but not a virtual certainty, a front end discount is imposed on the contractor's profit figure.

An example of a case where two of the three potential discounts were considered, but only one was taken, is found in the *Sound Contracting* case. Sound brought two bid claims against the City of Nainamo in the same action and was successful at trial in one of them. In the successful part of its case, the court implied a duty of fairness into Contract A, and held that the City had breached it. The City had used an apples-and-oranges analysis to compare two competing bids, and had effectively discriminated against Sound.

When the City argued that it should only have to pay the cost of bid preparation, the court demurred. But for the City's breach, would Sound have been awarded Contract B? The court answered this in the affirmative (no discount here).

The court then looked at Sound's claimed losses, which were a combination of lost mark-up plus items where revenues anticipated from the job would now not be received. The court dissected this claim on a line by line basis, and allowed the mark-up but discounted some of the other line items on the basis that they would be mitigated (avoided or reduced) by the contractor. In this case, the discount applied amounted to approximately one-third of the amount claimed. The *Sound Contracting* case was reversed on appeal. But the trial court's thinking on damages is a useful example of how a court may operate to arrive at a damage award.

Common sense tells us that a contractor that is rejected unjustly will behave the same way as a contractor whose bid simply fails. Both will look for other work (that is, mitigate/avoid losses). However, it is difficult to obtain meaningful evidence on what results mitigation should produce. The success of a contractor in finding alternative work is governed by a number of factors, many of which the contractor cannot control. Taking a discount from damages where a contractor has been unsuccessful in attempting to find replacement work seems to be a difficult thing for a court to quantify. Difficult or not, it is the duty of the contractor to try to mitigate its damages; the contractor that cannot demonstrate reasonable and good faith efforts in that direction will suffer erosion from its damage recovery.

*CanAmerican Auto Lease and Rental Ltd. v. R.* is another example of the thought process which a court employs when assessing damages for breach of contract, in this case Contract A. In *CanAmerican*, the court grappled with both the amount of damages and, more importantly, whether or not the different kinds of damages claimed were contemplated by the owner as flowing to the bidder if the owner breached Contract A.

In this non-construction case, the federal government called bids for car rental companies to obtain counter space at its airports. A company interested in bidding could qualify to bid for counter space that served either domestic or international arriving passengers. *CanAmerican* (really Hertz) qualified to bid for the international counter space as did its arch rival Avis (remember: Hertz was number 1, Avis was number 2 but tried harder). Neither Hertz nor Avis were interested in domestic counter space. Unfortunately, there were more international bidders than counters, and this complicated the bidding process.

Hertz developed a strategy which, if successful, would see it obtain a counter while its rival, Avis, would be put off site. Being off site in the rental car wars meant disaster.

The first component of the Hertz strategy lay in the bid documents. These documents stipulated that if a company bid for both a domestic counter and an international counter, only the higher bid of the two would be taken.

The second component in the Hertz strategy was that Tilden, a Canadian company, was qualified to bid on both counters. Hertz reasoned that Tilden's higher bid would be made for international space. If that happened, and given the owner's selection criteria, then if Hertz outbid Avis and Tilden, the limited number of counters would find Avis with no chair when the music stopped.

When the bids came in, the higher Tilden bid was for international space. Hertz succeeded in bidding higher than both Tilden and Avis. Tilden's bid for international space was higher than its bid in the domestic section. If the owner followed the bid documents, Avis would finish last and not get a counter.

At this point, the owner delivered a curve ball. If it took Tilden's higher international bid, it would lose Tilden domestically, *and* the next domestic bidder below Tilden was much less lucrative than Tilden. The owner, thus, rejected Tilden's international bid, and this gave Avis a counter in direct competition with Hertz.

The trial court held that the owner had breached Contract A by failing to follow through on its express bid selection criterion. The court had to determine what damages Hertz had suffered.

Hertz claimed damages on two fronts. First, had it known that the owner would behave as it did, it would not have bid as high for a less profitable asset. Its damages for overbidding were in the order of $250,000. Second, Hertz was now faced with a loss of business because its arch rival was right next to it. The trial court agreed that the overbid was a direct result of the owner's breach of Contract A. The court also awarded Hertz damages for the loss of profit due to competition from Avis.

On appeal, the Federal Court of Appeal upheld the breach of Contract A and agreed that Hertz should get damages for its overbid. On the loss of profit because of the presence of Avis, the appeal court hesitated. How much business would Hertz lose? Did it make any difference whether Avis was slightly up the road, or elbow to elbow? What damages were foreseeable to the owner? The court held that while the owner could foresee the overbid damages, it could not foresee the loss of profit.

The owner could not have foreseen that a loss would flow to Hertz because Avis was not forced off the airport. The appeal court had answered "no" to the question "whether in all the circumstances the loss or injury was of a type which the parties could reasonably be supposed to have in contemplation".

Does a car rental case teach any lessons to construction bidders? It should.

One lesson is that Contract A is not formed using a cookie cutter. The bid documents define Contract A, and a bidding authority cuts that cloth.

The *CanAmerican* case also shows how damages can be limited if the loss suffered by the wronged party was not in the "reasonable contemplation" of the party breaching the contract. Do not assume that the other party understands what damages you may suffer if there is a breach. Tell them up front!

## Awards In Favour of Subcontractors Against Contractors

As seen earlier in this book, Contract A between contractors and subcontractors is a different animal from the one between owners and contractors. Even so, a breach of Contract A results in damages flowing from contractors to subcontractors. And they are measured in essentially the same way as damages assessed against owners in favour of contractors.

By the time *Naylor* reached the Supreme Court of Canada, three different courts had provided three different calculations of damages — based on the same facts. Obviously, the calculation worth most note was made in the Supreme Court of Canada. But it is a worthwhile exercise to follow all three courts as they attempted to determine fair compensation following a breach of Contract A by Ellis-Don.

The *Naylor* case produced a surprising result at trial. Naylor had been the low electrical contractor in the bid depository and had been named and carried by the contractor in the contractor's prime bid. While the owner was considering what it would do, the contractor received an Ontario Labour Relations Board decision which confirmed that its union arrangements were in conflict with Naylor. Yet, when the contractor signed with the owner, Naylor was still the "carried" electrical trade.

The contractor did not advise Naylor that it had the job, and the contractor later replaced Naylor with another subtrade at the same price that Naylor bid. The trial judge held that Contract A between Naylor and the contractor had *not* come into being because there was no communication of acceptance. Troubled that Naylor would have no relief, the trial judge concluded that the contractor was *unjustly enriched* by using Naylor's price to secure another. He awarded Naylor $14,500 to compensate for its cost of bid preparation.

In March 1999, the Court of Appeal for Ontario reversed part of the trial court's decision in *Naylor*. The court held that Contract A between Naylor and the contractor did come into being when Naylor's name was "carried" in the prime bid as the electrical subtrade.

The trial judge had found that Contract A had not come into being between Naylor and Ellis-Don. Because trial judges have the benefit of hearing the evidence and assessing its worth, it often happens that a trial judge will provide his/her measure of damages even though he/she has found the damages are not payable. So it was in the *Naylor* case where the trial judge determined that, had Contract A been in existence and been breached, Naylor's damages would have been in the order of $730,000 — roughly one-half of the amount Naylor claimed. No doubt, the trial judge was influenced by the fact that the replacement contractor had suffered a significant loss carrying out the work.

The Court of Appeal concluded that the proper measure of damages for Naylor was its lost opportunity to have the subcontract. But the court felt it was appropriate to factor in some additional contingencies — contingencies the trial judge did not consider. First, because of the unfavourable labour ruling, Naylor might have had to share its profit with a subcontractor with the proper affiliation. As well, there were unforeseen problems on the job that were encountered by the replacement trade.

Ellis-Don argued for some additional deductions and that no effect should be given to a claim for profit on anticipated extras. The court rejected Ellis-Don's position but after considering the factors it felt appropriate, discounted profit by 50% for labour issues. The court then reduced the remaining balance a further 50% for unseen problems on the job. The award to Naylor, after all these discounts was $182,500 — one quarter of the trial judge's take.

At the Supreme Court of Canada, the finding that Contract A had come into existence was upheld. At the instance of Naylor, the Supreme Court of Canada examined the damage reduction which had been imposed by the Court of Appeal for Ontario.

The Supreme Court of Canada first addressed whether the appellate court should or should not have changed the trial judge's finding. The rule in such situations is that an appellate court should not substitute its own view unless the trial judge made an error in law or misunderstood the evidence in a fundamental way. Basically, the trial judge needs to have missed the point — on damages — significantly.

The Supreme Court of Canada had no difficulty with the appellate court's consideration of unanticipated site conditions. Here, the trial judge may not have fully appreciate the difficulty which the work presented due to a lack of as-built drawings and the need to perform the work within the functioning hospital. The Supreme Court of Canada allowed the appellate court's 50% discount to stand.

When the Supreme Court of Canada looked at the further discount taken by the appellate court in connection with labour difficulties, it found the appellate court's reasoning and appreciation of the trial evidence did not justify the second reduction. The Supreme Court of Canada fixed the damages at $365,143, half of the amount found by the trial judge, and twice the amount determined by the Court of Appeal.

## Awards In Favour of Contractors Against Subcontractors

Where a subtrade breaches Contract A, the measure of damages is essentially the same as when a contractor breaches Contract A with an owner. This measure is best illustrated by *Gloge*, one of the first cases to find that the Contract A/Contract B paradigm applies to relations between contractors and trades. It is also a good example of this damages model.

Gloge bid on the mechanical and electrical work to a company called Northern Construction. After Northern had carried Gloge in its prime bid, Gloge discovered that it had made an error and refused to execute the subcontract when Northern awarded it. Northern found a replacement trade at a cost of $340,000 above the Gloge bid.

The court held that Contract A obliged Gloge to perform the subcontract because it was awarded within the period of irrevocability. When Gloge refused, it breached Contract A.

The damages to Northern flowed logically from Gloge's refusal to do the work, and the full amount of Northern's loss (the spread between Gloge and the replacement trade) was awarded against Gloge.

## MISREPRESENTATION

### Awards In Favour of Subcontractors Against Owners

In rare circumstances, owners may be liable to subtrades. Because most bidding documents and most prime contracts provide that there is no contractual relationship between the owner and any subtrade, the Contract A/Contract B model does not apply. However, courts have another way of dealing with a tough-minded owner — the tort of misrepresentation.

In *Mawson Gage,* summarized earlier, Mawson submitted an electrical subtrade bid to the general contractor for some Government of Canada work in Ottawa. Quite by accident, the owner had omitted a section of the specifications from the set of bid documents which Mawson had picked up from the owner. After the bids were in and Mawson was carried by the low bidder, the mistake was discovered.

Mawson notified the contractor and the owner that there was an error in the bid documents, and Mawson requested a fix. The owner met this request by awarding the contract to the contractor, and the contractor in turn put the subcontract to Mawson. The subcontractor was thus forced to carry out the work, including the omitted section of the specification, and sued both the owner and the contractor.

The court held that the owner had a *duty of care* to fully represent the scope of work in all the bid documents which it issued. That duty was breached, and Mawson had suffered damages. Mawson was able to demonstrate that its loss was a direct result of its reliance on the owner's drawings and specifications (no owner mistake, no Mawson loss). The court was persuaded that to not give Mawson relief would give the owner something for nothing. Mawson was, therefore, awarded the value of the work that it was forced to do on the omitted section of the specifications.

In Chapter 5, we summarized two cases where subcontractors claimed unsuccessfully against owners (*Twin City* and *Ken Toby*).

In *Twin City*, the trial court held that the owner owed a duty of care to Twin City to police the behaviour of the contractor in relation to the subcontractor's bid to the contractor. The decision was reversed on appeal, the appellate court being satisfied that the owner was not obliged to look behind assurances which the contractor had given that all was well.

In *Ken Toby*, the subcontractor complained that the actions taken by the owner in the manner in which it made amendments to the bid documents both created Contract A between the owner and the subcontractor (a very unusual finding), and held that the owner had breached the duty of care.

On appeal, the court concluded no contractual relationship had been created between the owner and the subcontractor when the subcontractor bid to the bid depository. As to a duty of care, the court held that the owner had acted reasonably in the way it had conducted its process.

In both *Twin City* and *Ken Toby*, the appellate court found that the behaviour of the owner was reasonable in relation to the complaining subtrade. Put another way, the owner was expected to behave to a standard of care which each court defined. In both cases, the owner has met the standard.

It would be unwise to conclude from the appeal court decisions in *Twin City* and *Ken Toby* that an owner owes no duty of care to the subcontractors which animate the owner's bid process.

## Contractors Against Owners

As stated earlier in this Chapter, the measure of damages for tort (misrepresentation) is that the claimant is put in the same position it would have occupied had the misrepresentation not been made. If a contractor has no remedy under Contract A, the remedy in misrepresentation may cover a contractor's actual loss (*i.e.*, its cost of bid preparation or the cost of work done pursuant to the misrepresentation). But it will not compensate the contractor by paying it the profit it might have earned on work it might have done. Because lawsuits should (normally) be conducted with an eye on value (recovery) for money (the cost of litigation), the two cases need to be illustrated.

In *BG Checo*, the owner misrepresented the state of a power line right-of-way. The owner had undertaken to clear it, but didn't. BG Checo suffered the cost of clearing the right-of-way that it was told was not in its scope of work. As a result of the owner's misrepresentation, BG Checo suffered the actual cost of clearing the right-of-way as part of the construction contract (Contract B). Had the misrepresentation not been made, the right-of-way would have been cleared by the owner. Alternatively, BG Checo would have included the cost of clearing the right-of-way in its bid.

A similar illustration is found in the *Brown & Huston* case. There, the owner's consultant omitted important soils information from the site data made available to all contractors. The omitted information dealt with very difficult subsurface conditions which Brown & Huston encountered and had to overcome, at significant cost. Brown & Huston succeeded in its claim against both the consultant and the owner for misrepresentation. Because Brown & Huston contributed to its own loss by failing to make sufficient inquiries on its own, its recovery was reduced by 25%. But the owner and the consultant were jointly and severally liable to Brown & Huston for damages suffered because of the misrepresentation about working conditions.

> **Joint and Several**. A legal expression meaning "together and separately". If two or more persons enter into an obligation that is joint and several, their liability for its breach can be enforced against them all or against any of them separately.

In a bid case where the claimant has not been awarded Contract B, the pickings are much slimmer. A claimant must make out the five elements of misrepresentation (see *Hedley Byrne* summarized earlier in Chapter 9). The actual loss of the bidder is (usually) the cost it incurred to prepare the bid. That level of damage award is true to the theory of damages in tort — to put the party in the position they would have occupied had the misrepresentation not been made in the first place. So, our wronged bidder, had it known the true facts, would not have bid.

On a cost/benefit analysis, a misrepresentation suit by a bidder, which was not awarded Contract B, will usually be a losing proposition.

## OTHER RELIEF

### Unjust Enrichment

The trial judge in the *Naylor* case awarded $14,500 in damages for *unjust enrichment*. He found that the contractor had indulged in bid shopping which, while not illegal, was certainly antisocial.

In the Court of Appeal for Ontario, the court held that the contractor had breached Contract A, and was liable to Naylor in damages to the tune of $182,000. In the course of its reasons, the Court of Appeal took a brief look at the finding of the trial judge respecting unjust enrichment. This passage was included in the reasons of Justice Weiler speaking for the Court of Appeal:

> I would also find that Ellis-Don was unjustly enriched at Naylor's expense. It is more appropriate, however, to assess the damages on the basis that Ellis-Don breached the terms of its preliminary contract A with Naylor.

When *Naylor* went to the Supreme Court of Canada, it barely touched on the notion of unjust enrichment. That Court's take on this form of relief was that "there is no need to examine the alternative ground of unjust enrichment relied upon by the trial judge."

The fact that unjust enrichment may be a remedy available to a subtrade which has had its price shopped puts some frosting on the *Naylor* cake. Not only did the Court of Appeal decision in that case spell out a tight, subtrade-friendly Contract A, it gave those same trades a back-up remedy should Contract A not be available. Based on the findings of the Court of Appeal, Naylor could have been compensated either for breach of Contract A or on the theory of unjust enrichment. The court proceeded on the theory of damages which gave Naylor the better return.

Unjust enrichment is an equitable remedy. It is given in circumstances where somebody gets something for nothing, and there is no contract between the parties which can be used to cause the benefiting party to pay the other for the benefit. Since the Court of Appeal for Ontario endorsed unjust enrichment as a remedy in a bid setting, don't be surprised to see the claim again.

*Mawson Gage* was a negligence case, but it could also have proceeded on a theory of unjust enrichment. The Government of Canada had to pay Mawson the fair value of the work the Government would otherwise have received (unjustly for free).

The trial decision in *Naylor* suggests that the calculation of damages for unjust enrichment may not always be simple. The trial judge awarded Naylor $14,500 because he felt that the work that Naylor had put into its bid helped the contractor find a replacement trade at the same price. The cost of Naylor's bid preparation was an odd way to evaluate that benefit, but it seems to be all that was available.

When damages are difficult to prove, courts will usually proceed on a "best evidence" basis. The fact that a loss cannot be proven in minute detail does not mean that there was no compensable loss. Nor does it mean that a court will deny compensation for it. The closer the damage evidence comes to science than art, the better a party's chances for a meaningful damage recovery.

## Prerogative Remedies/Injunction

The *Thomas C. Assaly* case was decided in the Federal Court of Canada in 1990. Assaly and three others had responded to a federal government bid call for office space for the new GST agency. The Instructions to Bidders failed to specify that the federal government wanted the agency in a building that was free from other tenants. The two low bidders were passed over (Assaly being one of them) because the owner went for the bid which, quite by accident, made the agency the sole tenant. Assaly sought an order to "quash" the owner's decision on the basis of procedural unfairness. The court agreed with Assaly's assertions and directed the owner to reconsider its decision. In effect, the court was ordering a "fair procedure" without telling the owner what that "fair procedure" should produce as a decision.

In the *Assaly* case, the minister decided to call new bids shortly after the court ordered a reconsideration. The new bid process resulted in further court applications by the bidders, in one case to quash a step taken by the owner and in the other to force a step that was not taken.

The application to quash, known as *certiorari*, and the application to force action, known as *mandamus*, are called *prerogative remedies*. They are available in circumstances where a party, usually government or a government agency, has a statutory or regulatory obligation to act and either refuses to act (*mandamus* — to force action) or acts in excess of its statutory or regulatory authority (*certiorari* — to undo action).

There are other case summaries where these prerogative remedies are invoked by contractors that feel aggrieved in a bid process with government. If these remedies are granted, government actions taken may be undone, or actions not taken, started. But damages are not awarded.

The compel/undo remedies can cause or rewind administrative actions but cannot direct their outcome. The fact that an agency acts fairly does not necessarily mean that the contractor will get what it wants. In some of the cases summarized (see *Ellis-Don Construction Ltd. v. Canada* and *Peter Kiewit*), the prerogative remedies were refused, and the court rightly directed the complainants toward the usual damage remedies sought in court.

The prerogative remedies remind one of how the garden variety privilege clause works. Contractors often read the privilege clause and believe that it means that, if all the bid criteria are applied fairly and not breached, the lowest bidder will get the work. It is not necessarily so. The owner may be obliged to reject another bidder, but the owner is usually under no Contract A obligation to accept the bid of the complaining contractor. By the same token, a prerogative remedy may require a fair process or stop an unfair one, but it cannot directly deliver Contract B.

The *Donohue* decision, reviewed earlier in this book, is another example of a non-compensatory remedy. In *Donohue*, the presiding judge granted injunctive relief in a procurement case — one where the process was labelled "RFP."

Presiding over an appliation under the *Rules of Civil Procedure* for Ontario, the judge ordered the City of Toronto to reconsider Donohue's submission in response to the RFP. But there is no certainty that the City's second assessment of Donohue's submission will go better than the first.

The subtext in the *Donohue* decision — for what it's worth — is that the court treated Contract A like any other contract, and found the City had a contractual obligation which it had not discharged.

## AFTERTHOUGHT

Like the rest of this book, this Chapter is not intended to make the reader an expert in damage theory or alternative remedies. Our hope is that by outlining where litigation leads, in a best case scenario, our readers will have one more piece of information available when making the "go/no go" decision on bid litigation.

# Chapter 14

# CONTRACT A: TREAT IT LIKE A CONTRACT

## CONTRACT A IS A REAL CONTRACT

A bit of judicial conjuring in the *Ron Engineering* case created Contract A — the bidding contract — and then labelled the construction contract Contract B. The case involved a mistake in bid pricing. The decision looked like a one-way street: the owner won, the contractor lost.

The subsequent case law has broadened the application of Contract A so that, today, it is *absolutely* clear that Contract A is a real contract in every sense of the word. It is also clear that Contract A can come into being between both the owner and contractor, and the contractor and its subtrades.

So, any of these parties can enter and breach Contract A.

Contract A is unusual. In most cases where a contract is made, one party makes an offer followed by the other party's communication of acceptance of that offer (while the offer is still open). A contract also requires *consideration*, usually where the parties assume mutual obligations (see, for example, *Dickinson v. Dodds* in Chapter 1 of this book).

In *Ron Engineering*, Justice Estey described Contract A as a "unilateral contract." Today, the dialogue around Contract A is directed to what the parties intended. Mostly, intention is deduced from the type of process that the bid documents describe. Chances are, the owner and bidder in *Ron Engineering* did not *intend* to enter Contract A or any type of "process contract." That may be why some judges, over the last 20-plus years were not prepared to accept that the act of submitting a compliant bid gives rise to a contract at all. Some in the judiciary and many in the industry viewed *Ron Engineering* as a bad dream, and hoped it would be gone when they awoke. No such luck.

The 1999 decision of the Supreme Court of Canada in *M.J.B.* was quite specific and, to the critics of *Ron Engineering*, disappointing. The Court held that the Contract A/Contract B model described in *Ron Engineering* exists in all its complexity. But Contract A is not automatic. In its very practical approach to the model, the Supreme Court held that the terms of Contract A, in any given case, depend on a specific bid package. This is both good sense and good law.

In 2000, the Supreme Court issued its ruling in *Martel*. In addition to further affirming the Contract A/Contract B model, the Court anointed the rulings of lower courts that an implied duty of fairness is a term of Contract A, binding the owner, unless specifically disclaimed.

The irony in *Ron Engineering, M.J.B., Martel,* and *Naylor* is that nowhere in the bidding documents before those courts was there an express offer of Contract A. Well, it is not the fault of the court which must deal with what is in front of it. Isn't it time, so many years after the first appearance of Contract A, that we *treat it like a contract*?

## TREAT CONTRACT A LIKE A CONTRACT

Some industry participants and industry watchers view the Supreme Court of Canada decision in *Ron Engineering* with dismay, thinking of Contract A as a drive-by shooting perpetrated on the law of mistake. After all, *Ron Engineering* flowered to address a chronic bid problem, namely how to reconcile an ostensibly "formal" irrevocable bid with the bidder's withdrawal on grounds that it made a serious mistake. Classic mistake theory meant that the owner could not accept a bid, even an irrevocable one with bid security, once advised of the mistake in the bid. Then came *Ron Engineering*.

The immediate reaction to *Ron Engineering* — apart from the shock of it all — was that contractors were walled off by Contract A from obtaining relief under the law of mistake for an invisible and serious bid error. But as time went by, the focus shifted from mistake to the other implications of Contract A. In a journey which has lasted more than two decades and stretched from coast to coast and back, Contract A has been elaborated to define the obligations of both bidder and owner. Owners discovered, to their surprise, that Contract A imposed obligations on them and, in some cases, very significant damages if those obligations were not met.

It is now settled that Contract A (usually) imposes an implied duty of fairness on the owner and (usually) places an obligation on the owner not to award Contract B to a non-compliant bidder.

No question about it, *Ron Engineering,* and the cases which have refined it, revolutionized the law of bidding.

### New Law/New Practice

One would think that the evolution of Contract A would spawn corresponding changes in the terms of bid documents. Wrong! At least, largely wrong.

Few owners issue bid documents (whether called bid documents or procurement documents labelled "request for proposal") which speak of creating the process contract which is Contract A.

More fundamental, the industry still seems to view Contract A as a *weapon*, typically drawn during a bid dispute, rather than as a tool framing the bid process and controlling its outcomes.

Consider a typical construction contract such as CCDC-2 1994. The drafters of that contract took stock of the risks and obligations involved in executing a project. They set out the commercial terms: what is to be built, when, for how much. They also listed a series of obligations and identified contract risks. Then they allocated the risks between the parties, or left them at large if that was the better decision, and then managed or insured them.

Why does the construction industry not take the same approach to Contract A? It's a mystery.

If typical bid documents expressly offer to enter Contract A with compliant bidders, a number of good things may happen.

First, both parties know that contract obligations are in the offing and that value is exchanged for the assumption and discharge of obligations.

Second, a court or arbitrator would not be confused about whether this particular set of bid documents was, or was not, intended to create Contract A (always subject to the "duck test").

## Identifying and Allocating Risks

Most sophisticated commercial contracts anticipate a breach of that contract and how the parties will behave if a breach occurs.

In Contract A, one of the principal risks is that the bidder will refuse an award of Contract B. Another risk is that, inadvertently or otherwise, the owner will breach its Contract A evaluation/award obligations.

Under most bid documents, a low compliant bidder who refuses to accept an award will be liable for the difference between its bid and the second lowest compliant bid. This is, potentially, a large amount of money, especially if a mistake has been made. Contracting being what it is, a substantial damage award can wipe out all but the most robust of contractors.

On the owner's side, a breach of Contract A can bring damages which range from the nominal (the cost of bid preparation) to the significant (the discounted value of the contractor's anticipated earnings for the project).

Considering the nature and purpose of the bid exercise, the punishment may often be much too harsh for the crime.

Consider a bid document, which identifies Contract A as the "process contract" it is, and includes clauses like the following:

*1.1     The Bidder acknowledges that by submitting a compliant bid, it has accepted an offer by the Owner to enter a "process contract" for the evaluation of bids and the award of the Contract, if an award is made. The Bidder acknowledges that the terms of the "process contract" are described in the Bid Documents.*

*1.2     The liability of the Bidder to the Owner for loss and damage arising out of the Bidder's breach of the "process contract" shall be limited to the lesser of the actual loss suffered by the Owner and the sum of One Hundred Thousand ($100,000.00) Dollars.*

*1.3     The liability of the Owner to any Bidder for the negligence of the Owner or for the breach by the Owner of the "process contract" shall be limited to the lesser of the sum of One Hundred Thousand ($100,000.00) Dollars and the reasonable cost to the Bidder of preparing its bid.*

These clauses have not had the approval of a court yet. And, they are only an illustration. But, Contract A being a contract, the mutual limitation of liability should be effective if it is clear enough.

If this limitation clause, or one like it, is included in bid documents, it affects bid security. If the maximum damage for which the bidder can be liable for a breach of Contract A is $100,000, there is no point in requiring bid security which exceeds that amount. This may or may not produce savings in bid prices. It will help to reduce bid disputes. Properly set, the limitation will not cause irreparable damage to a contractor if it makes a bid mistake or refuses an award.

The above example arbitrarily uses $100,000 as the limit of liability but the limit has to be assessed for each bid process. A balance should be struck between a number that is high enough to discourage anti-social bidder behaviour, but not so high as to inflict ruin. The corollary is that owners, by accepting these limits, give up the opportunity to reap a windfall from a bidder's breach of Contract A. This is as it should be.

## Keep It Simple — Keep Your Options Open

In Chapter 6, we looked at the bid documents in the *Toronto Transit Commission v. Gottardo* case. There, the owner, perhaps inadvertently, created a two-stage bid process, even though there was no reference to two closings in the bid documents. The TTC created a definition of bid documents, which was very broad. It included the usual components: instructions to bidders, bid form, contract documents. It went on to list within the definition a volume of supplementary information which the bidder was to provide only if given notice by the owner. When the literal terms of the definition were identified, it turned out that the initial bid submission by any bidder was only part of the story.

So, if one of the purposes of Contract A is to create a term of irrevocability binding the bidder, the owner in *Gottardo* left the barn door open and the horse free to leave. Because, when Gottardo's bid was opened, it immediately realized that the bid included a serious pricing error — one invisible to the owner. Showing no sympathy for Gottardo's plight, the owner attempted to invoke its Contract A right to hold Gottardo to its bid. As it moved in that direction, the owner gave Gottardo notice that it was required to submit the second stage of information to complete the bid documents. When Gottardo balked, the owner found itself hoisted on its own very broad definition of bid documents.

In the *Gottardo* case, whatever the owner's intentions, the elaborate definition of bid documents thwarted the process. The Gottardo bid was non-compliant with the owner's definition of bid documents. In all likelihood, the Court of Appeal will leave the decision alone.

In some ways, it is unfair to make an example of the *Gottardo* bid documents. Owners everywhere are waxing elaborate when simple might serve them better. *Gottardo* represents a useful and recent illustration of how an attempt to create additional Contract A obligations acted to frustrate rather than foster the submission of compliant bids.

In any contract, the word "shall" is very powerful. It is also limiting. A party subject to "shall" obligation either performs it, or finds itself in breach.

By contrast, presence of the word "may" in contract sends a different signal: the party has options.

In a bid setting, the choice of "shall" or "may" when applied to a matter of compliance or evaluation should be the subject of careful decision-making. The decision is easy around issues such as failure to submit a bid on time or failure to submit the required bid security — "the bid *shall* be ruled non-compliant". But, elsewhere in the bid documents the use of "shall" rather than "may" — or vice versa — may change the entire outcome of a bid process.

For example, the bid documents state that the listing of trades is mandatory. The same provision may go on to say that a failure to list trades "may result in the bid being ruled non-compliant". Shouldn't it be "shall"? Not necessarily.

If the owner in the above example uses "shall" on a contract then a bidder who omits a very minor trade must be ruled non-compliant. Failure by the owner to make such a ruling would be a breach of Contract A with compliant bidders. But if the word "may" has been selected, the owner has a choice. If the omission is minor, it can elect not to disqualify. If an owner wants to have a broad scope to determine substantial compliance, "may" is the tool.

As another aspect of life, too much of a good thing can be bad. Owners that use "may" for things like late bid delivery and failure to deliver bid security don't do themselves any favours. If the language of the bid documents is so loose and so open, it may wind up destroying the very essence — Contract A — that the bid was intended to create. Little words — "shall" and "may" — but important.

So, as we explore how to treat Contract A like a contract, perhaps we might consider a move towards bid simplicity, a move back to the basics.

## Dispute Resolution

Over the last ten years, a great deal of effort has been devoted to alternative dispute resolution (ADR), meaning alternative to litigation. In Ontario, the court process now includes mandatory mediation in all but a few types of case bound for court. Most construction contracts, CCDC-2 1994 being a prime example, include a negotiation/mediation/arbitration mechanism intended to keep construction disputes out of court and in the hands of persons who understand construction and its culture.

Contract A is not a one-on-one situation. In larger bid calls, there can be five or more bidders, all of whom may have Contract A with the bidding authority. If you use regular ADR like the CCDC-2 model in such a bid context, how do you avoid creating an ADR circus?

With multiple process contracts and an industry populated by people who are not shrinking violets, garden-variety dispute resolution provisions don't work, at least they don't work for the owners who control the contents of bid documents. So, any dispute resolution provision must be attractive to owners.

Consider a dispute resolution provision such as this:

> 1. In the event of a dispute arising in connection with this bid process, including a dispute as to whether the bid of any Bidder was submitted on time or whether a bid is compliant, the Owner, in its unqualified subjective discretion, may refer the dispute to a confidential arbitration before a single arbitrator with knowledge of procurement/bidding law and practice at Toronto, Ontario, pursuant to the Arbitration Act, 1991, as amended. In the event that the Owner refers the dispute to arbitration, the Bidder agrees that it is bound to arbitrate such dispute with the Owner. Unless the Owner shall refer such dispute to arbitration, there shall be no arbitration of such dispute.

> 2. In the event the Owner refers the dispute to arbitration, the Owner and the Bidder agree that they shall exchange brief statements of their respective positions on the dispute, together with the relevant documents, and submit to an arbitration hearing which shall last no longer than two (2) days, subject to the discretion of the arbitrator to increase such time. The parties further agree that there shall be no appeal from the arbitrator's award.

Some will argue that this clause is not fair because it is not symmetrical (only the owner can trigger an arbitration). It is not symmetrical because Contract A is not symmetrical either: one owner/several bidders. Besides, the damages that a bidder may recover in court for a breach of Contract A fall off dramatically if the complaining bidder is not the lowest compliant bidder. The compliant bidder that finishes third, fourth or tenth in the bid process may recover only nominal damages, perhaps the cost of bid preparation, because it never really had a chance — breach or no breach of Contract A. So, this dispute clause reflects "process contract" dynamics and a means to persuade the owner to include the clause in its bid documents.

Unfair? Not at all. Any bidder still has its civil right to bring a lawsuit. If the damages at stake are minimal, contractors, being very smart people, will probably pass.

### SO, WHERE ARE WE?

Most bid documents are prepared by consultants and their specification writers, often without lawyer input on the "process contract" issues. This may be why owners rarely use Contract A as a tool for allocating, managing and reducing at least some of the risks in bidding.

# Chapter 15

# CONCLUSION

It is not often that one gets an opportunity to witness a true revolution right from the start, and document its progress. *Ron Engineering* indeed represented a revolution, and in this book we tried to be its historians and commentators. We also tried to offer practical advice on how to survive and, if possible, prosper in these interesting times.

Twenty-three years after the outbreak of the unrest, it is tempting to say that things are finally settling down. However, every time we get this feeling something new comes up thanks to the ingenuity of the industry participants, their lawyers and our courts. Some important issues still remain to be settled by the Supreme Court of Canada, probably a few years from now.

Has the revolution been good for the construction industry?

Undeniably, it has shaken up the industry and made everyone more aware of the good and bad practices associated with bidding. As a result, the process is probably a bit more fair than it used to be. It is certainly far more risky, with potentially devastating penalties for bidders and owners who transgress against the new rules, penalties totally out of proportion with the transgressions.

Is it really a benefit to the construction industry that a contractor or an owner may now be hit with damages amounting to hundreds of thousands of dollars for a mistake in putting together a bid, or in assessing bid compliance? Have the hundreds of court decisions since *Ron Engineering* and thousands of hours of consultation with lawyers really brought enhanced certainty for bidders, as claimed by the judges, or is the opposite true?

Probably the most lasting and indisputable effect of the revolution is that lawyers and courts are now an integral part of the bidding process, transferring millions of dollars out of construction into litigation.

Maybe that is simply the right price for the construction industry to pay for its continuing inability to put its own house in order, using its own resources.